Anammox Technology in Industrial Wastewater Treatment

Maulin P. Shah
Editor

Anammox Technology in Industrial Wastewater Treatment

Springer

Editor
Maulin P. Shah
Environmental Microbiology Lab
Gujarat, India

ISBN 978-981-99-3458-4 ISBN 978-981-99-3459-1 (eBook)
https://doi.org/10.1007/978-981-99-3459-1

This Springer imprint is published by the registered company Springer Nature Singapore Pte Ltd.
The registered company address is: 152 Beach Road, #21-01/04 Gateway East, Singapore 189721, Singapore

Contents

Ammonia-Oxidizing Bacteria in Wastewater

Ankita Chatterjee and Maulin P. Shah

Abstract Ammonia-oxidizing bacteria (AOB) are important for generating nitrite from ammonia in the wastewater. The initial step involved in nitrification is ammonium oxidation resulting in nitrite. This step is carried out by ammonia-oxidizing bacteria. Ammonia-oxidizing bacteria are classified under the class of Proteobacteria, most of which are beta and gamma-proteobacteria. There are several parameters affecting the activity of ammonia-oxidizing bacteria, which are, presence of free ammonia, regulation of the dissolved oxygen concentration, pH, temperature, alkalinity, solid retention time and hydraulic retention time. The effect of these parameters on ammonia-oxidizing bacteria would be discussed in the chapter in detail. Ammonia-oxidizing bacteria are able to degrade micropollutants present in wastewater. The nitrogen constituents present in wastewater are transformed from ammonia to nitrate through nitrification using the ammonia-oxidizing bacteria. In this chapter, the application of ammonia-oxidizing bacteria in the biodegradation of micropollutants is discussed in detail.

Keywords Ammonia-Oxidizing Bacteria (AOB) · Biodegradation · Nitrite · Nitrification · Micropollutants

1 Introduction

The abundance of nitrogen in the ecosystem is circulated among plants and animals via the nitrogen cycle. Interference of human activities in the natural ecosystem, however, causes an imbalance in successful accomplishment of nitrogen cycle and thus affects the soil, water and air. Nitrogenous compounds, when released in excess from the sewage plants to large water bodies, result in uncontrolled growth of algae in

A. Chatterjee (✉)
Department of Biotechnology, School of Applied Sciences, REVA University, Bangalore, Karnataka, India
e-mail: ankita.chatterjee@reva.edu.in

M. P. Shah
Enviro Technology Limited, Ankleshwar, India

M. P. Shah (ed.), *Anammox Technology in Industrial Wastewater Treatment*,
https://doi.org/10.1007/978-981-99-3459-1_1

the ponds, lakes or sea thereby reducing the dissolved oxygen level. This initiates the process of eutrophication and eventually causes the death of aquatic lives. Thus, the process of reduction of nutrients in the wastewater sewage effluents must be effective. Elimination of nitrogen content by biological process is widely accepted across the globe. The process involves the application of nitrification and denitrification. Microorganisms play a crucial role in completion of the biological mechanisms. Biological nitrogen fixation includes reduction of unreactive molecular nitrogen to compounds of ammonium (Fowler et al., 2013). Biological oxidation mechanism is involved in nitrification where aerobic autotrophs play a major role. Ammonia-oxidizing bacteria (AOB) oxidize the ammonium to produce nitrite during the first phase of nitrification followed by the activity of nitrite-oxidizing bacteria in the second stage where the nitrite is further oxidized to nitrate. On the contrary, the nitrates are reduced to nitrite by the process of denitrification. Reduction of the nitrites is further carried out by anaerobic heterotrophic microbes to result in the formation of nitrogen gas, thus completing the nitrogen cycle. However, the challenges faced during the successful completion of biological nitrogen removal, such as delayed growth of the important microorganism associated with the techniques, increased oxygen and carbon demand as well as to keep control on the continuous nitrification and denitrification factors, hinders the process to a certain extent. Thus, the process of shortcut biological nitrogen removal was introduced in recent years where direct conversion of the nitrites to nitrogen is conducted by anaerobic ammonia-oxidizing bacteria or heterotrophic bacteria. These microbes utilize the nitrites as electron acceptor (Soliman & Eldyasti, 2018). Aggregation of ammonia-oxidizing bacteria enhances the rate of the shortcut biological nitrogen removal process. In this process, the activity of nitrite-oxidizing bacteria (conversion of nitrite to nitrate) is suppressed or inhibited, however, no inhibitory regulations are implied on the activity of AOB.

2 Microbial Characteristics of Ammonia-Oxidizing Bacteria

The extended application of AOB ranging from the accomplishment of nitrogen cycle to actively participating in wastewater treatment to reduce ammonia content has intrigued the researchers to study the organisms in detail. The first isolation of AOB was reported around the nineteenth century and since then the diversity of AOB is explored. However, following extensive researches, isolation and identification of five genera of AOB were recognized and categorized within the class of Proteobacteria. According to the 16S rRNA sequencing analyses of the isolated AOB, among the five genera, four of them, namely, *Nitrosomonas*, *Nitrosospira*, *Nitrosovibrio* and *Nitrosolobus* belong to the subclass β-proteobacteria and *Nitrosococcus* is classified with γ-proteobacteria (Koops & Pommerening-Röser, 2001). The in-depth detailed study of AOB was challenging in the initial phase following the conventional enrichment method of isolation and identification. The introduction of molecular aspects, on the other hand, played an important role in rapid detailed study of the AOB. Culture-independent processes have been developed based on the already existing 16S rRNA sequences. The culture-independent process aids in exploring the AOB diversity and their habitats along with the growth type of these microbes in complex or unfavorable environments (Purkhold et al., 2000; Shah, 2020, 2021).

Morphologically, AOB can be studied based on their cell size, shape, presence of flagella and intracytoplasmic membranes. Table 1 shows the cell size and shape along with flagellar presence of certain AOB.

3 Factor Affecting the Activity of Ammonia-Oxidizing Bacteria in Wastewater

Successful accumulation of ammonia-oxidizing microorganisms aids in efficient nitrification. There are certain external factors that play important roles in the growth and mechanism of AOB. The responsible factors and their effects on the AOB are discussed in brief below:

- Temperature: Temperature directly affects the activity of AOB. It has been reported by Kim and Lee in an experimental setup conducted in 2011, when the temperature of the reactor was above 25 °C, removal of ammonia was more than 80%. However, at 15 °C, only 30% or below ammonia were removed. Nitrification activity of AOB increases by around 15–18% with the elevation of temperature (Zhang et al., 2015). Maintaining the temperature within 20–25 °C activates the ammonia oxidation efficacy as well as inhibits the oxidation of nitrite.
- Concentration of Dissolved Oxygen: Maintenance of the level of dissolved oxygen in bioreactor enhances the accumulation of nitrite and thereby accumulates AOB. Lower concentration of dissolved oxygen limits the growth of nitrite-oxidizing bacteria but the growth of AOB is not affected. This might be possible because of the higher affinity of AOB towards oxygen. One of the early reports regarding growth of AOB reveals that declined dissolved oxygen concentration doubles the growth rate of AOB (Hanaki et al., 1990). Limited concentrations of dissolved oxygen, in some cases, lower the rate of nitrification due to bulk deposition of sludge. In suspended growth conditions in activated sludge, it has been observed that maintaining the concentration of dissolved oxygen within 0.4–0.5 mg/L helps in accumulation of nitrite at a rate of 96% along with increasing the count of AOB cells (Mohammed et al., 2014). It has also been reported that elevated concentration of dissolved oxygen might affect the nitrite accumulation in a negative way (Wei et al., 2014).
- pH, Free nitrous acid and Free ammonia: pH affects the equilibrium of free nitrous acid and free ammonia showing an inhibitory activity of AOB and nitrite-oxidizing bacteria. It has been reported that at a concentration of 0.1–1 mg free ammonia per liter, nitrite-oxidizing bacteria show high sensitivity and their actions are inhibited. In order to inhibit the activities of AOB, the concentration of free ammonia should

Table 1 Morphological Characteristics of Ammonia-Oxidizing Bacteria

Class	Sub-class	Genus	Species	Cell size	Cell shape	Flagellar presence	Gram characteristics
Proteobacteria	β-Proteobacteria	*Nitrosomona-s*	*N. europaea*	(0.8–1.1 × 1–1.7) μm	Rods	Flagella present in the polar regions, some strains are non-flagellated	Gram negative
			N. eutropha	(1–1.3 × 1.6–2.3) μm	Rod to pear shaped	Flagellated, Motile	Gram negative
			N. halophila	(1.1–1.5 × 1.5–2.2) μm	Coccus	Tuft of Flagella, Motile	Gram negative
			N. communis	(1–1.4 × 1.4–2.2) μm	Rods	Tuft of Flagella, Motile	Gram negative
			N. nitrosa	(1.3–1.5 × 1.4–2.2) μm	Pleomorphic (Rods or coccus)	Tuft of Flagella	Gram negative
			N. oligotropha	(0.8–1.2 × 1.1–2.4) μm	Pleomorphic (Rods or coccus)	Polar Flagellum, Motile	Gram negative
			N. ureae	(0.8–1.2 × 1.1–2.4) μm	Pleomorphic (Rods or coccus)	Not reported	Gram negative
			N. marina	(0.7–0.9 × 1.7–2.2) μm	Rods	Not reported	Gram negative

(continued)

Table 1 (continued)

Class	Sub-class	Genus	Species	Cell size	Cell shape	Flagellar presence	Gram characteristics
			N. aestuarii	(1–1.3 × 1.4–2) µm	Rods	Not reported	Gram negative
			N. cryotolerans	(2–4 × 1.2–2.2) µm	Rods	Not reported	Gram negative
		Nitrosococc-us	*N. mobilis*	(1.5–1.7 × 1.5-2.1) µm	Rods or coccoid	Flagellated	Gram negative
		Nitrosospira	*N. briensis*	(0.3–0.8 × 1.0-8.0) µm	Vibrio or closely packed rods	Flagellated, Peritrichous, Motile	Gram negative
		Nitrosovibrio	*N. tenuis*	(0.3–0.4 × 1.1-3.0) µm	Curved rods	Lateral or subpolar flagella, motile or non-motile	Gram negative
		Nitrosolobus	*N. multiformis*	(1.0–1.5 × 1.0-2.5) µm	Pleomorphic	Flagellated, peritrichous, Motile	Gram negative
	γ -Proteobacteria	*Nitrosococcus*	*N. oceani*	(1.5–1.8 × 1.7-2.5) µm	Spherical or oval	Polar flagella, Motile	Gram negative
			N. halophilus	(1.5–1.8 × 1.7-2.5) µm	Spherical or oval	Polar flagella, Motile	Gram negative

be maintained at 10 mg/L (Anthonisen et al., 1976). The inter-relationship between pH and free ammonia (FA) can be depicted by the following equation:

$$FA(\frac{mg}{L}) = \frac{17}{14} \times \frac{NH_3 - N \times 10^{pH}}{10^{pH} + \exp(\frac{6344}{273+T}\}}$$

The relationship between pH and free nitrous acid (FNA) is represented by the equation mentioned below:

$$FNA(\frac{mg}{L}) = \frac{46}{14} \times \frac{NO_2 - N}{10^{pH} \times \exp(-\frac{2300}{273+T}\}}$$

4 Role of Ammonia-Oxidizing Bacteria in Degradation of Micropollutants in Wastewater

Existence of organic micropollutants, especially in the aquatic ecosystem, are one of the emerging concerns to human and animal survival. The micropollutants range from pharmaceutical residues to chemical sludges and chemical fertilizers and pesticides. Wastewater treatment plants are one of the major strategies followed these days for the treatment of the micropollutants. The molecular properties of the organic micropollutants play a specific role in determining the chances of the pollutant to undergo biodegradation. Thus, analyzing the molecular properties of the organic compounds before initiating the wastewater treatment process is much required (Su et al., 2021). The AOB majorly follows two steps, that is, metabolism and cometabolism during the mineralization of micropollutants. Metabolism involves utilization of the micropollutants as important carbon and energy sources however, in the phenomenon of cometabolism, the organic compound does not play a role in growth of microbes but they are transformed from the parent compound to simpler non-toxic metabolites.

Ammonia-oxidizing bacteria, in recent days, are used in wastewater treatment strategies even in artificial environmental conditions, such as constructed wetlands, biofiltration unit, drinking water treatment plants and many more. Few species of *Nitrosomonas*, like, *N. oligotropha*, *N. europaea* and *N. nitrosa* have been reported to be predominant AOB present in the domestic wastewater treatment facilities (Zhang et al., 2015). *Nitrosomonas* dominance in municipality wastewater treatment plants in China and Thailand has been reported in various studies interpreting that the species belonging to this genus are effective in treating or eliminating the pollutants from the wastewater (Gao et al., 2013; Limpiyakorn et al., 2011; Wang et al., 2012).

AOB secretes the enzyme ammonia monooxygenase (AMO), which acts as the functional enzyme for conversion of ammonium to hydroxylamine. Further, the hydroxylamine is transformed into NO_2 with the help of enzyme hydroxylamine oxidoreductase. The unique hydroxylamine oxidoreductase enzyme, present in the periplasmic space of AOB, aids in the second step of nitrification where oxidation process of hydroxylamine to result in NO_2 formation. This step is conducted using water as the source of oxygen (Kumwimba & Meng, 2019). The ammonia monooxygenase contains trinuclear copper centers at active sites along with iron ions (Gilch et al., 2010). Three protein subunits encoded by amoA, amoB and amoC genes in the AMO enzyme are hypothesized to play important roles in functioning of the enzyme. A portion of amoA acts as a key gene for catalyzing the initial reaction of ammonia oxidation (Arp et al., 2002). Thus, the diversity of ammonia-oxidizing bacteria is studied using the amoA gene as biomarker.

Ammonia-oxidizing bacteria follow cometabolic pathway, that is, the enzymes secreted by the AOB biodegrade the pollutants. AOB are also to degrade a range of micropollutants, ranging from alkenes, aromatic hydrocarbons, halogenated hydrocarbons, methane to several organic micropollutants (Xu et al., 2016). Few studies carried out on biodegradation of certain pollutants in wastewater using AOB are listed in Table 2. Remediation of pharmaceutical compounds using AOB is widely accepted in recent days to curb the immense toxicity of spreading pharmaceutical component in the environment. Residues of pharmaceutical compounds in lower concentrations are not utilized by the microbes as their sole carbon or energy sources. However, the ammonia-oxidizing bacteria efficiently degraded the pharmaceutical compounds into simpler metabolites via cometabolic biodegradation as shown in Fig. 1 (Ge et al., 2014; Xu et al., 2016). Dawas-Massalha et al. (2014) reported that various pharmaceutical compounds, such as carbamazepine, dexamethasone, ibuprofen, iopromide and ketoprofen, in trace concentration can be completely degraded by AOB. During the biodegradation process, a constant AMO activity was recorded. Earlier, in 2009, Roh et al. experimented the application of single strain of AOB, *Nitrosomonas europaea*, in biodegradation of triclosan, bisphenol A and ibuprofen. Though triclosan and bisphenol A were degraded by the strain, ibuprofen concentration did not show any change. However, the ibuprofen was efficiently mineralized by mixed AOB cultures.

5 Future Scope and Opportunities of Ammonia-Oxidizing Bacteria

In partial nitrification conditions, the accumulated ammonia-oxidizing bacteria act as important energy source and also for recovering or eliminating resources from the wastewater treatment plants. The ammonia monooxygenase enzyme of AOB aids in production of methanol by oxidation of methane. The methane-oxidizing bacteria that are responsible for producing methanol by methane oxidation, secrete enzyme called as methane monooxygenase. The ammonia monooxygenase present in the AOB shows much similarity to the methane monooxygenase and, thus, is able to produce methanol following the similar process as methane-oxidizing bacteria (Jones & Morita, 1983). Following this concept, AOB can further be used to prepare

Table 2 Biodegradation capacity of AOB in wastewater

Serial number	Name of the AOB	Degraded pollutants	Biodegradation rate	Degradation percentage	References
1	*Nitrosomonas europaea*	Estrone (E1), 17β–estradiol (E2), estriol (E3), 17α ethynylestradiol (EE2)	E1: 0.056 h^{-1} E2: 1.3 h^{-1} E3: 0.030 h^{-1} EE2: 0.035 h^{-1}	–	Shi et al. (2004)
2	Mixed culture of AOB	Sulfamethoxazole	–	86%	Kassotaki et al. (2016)
3	*Nitrosomonas europaea*	Bisphenol A	–	77%	Sun et al. (2012)
4	Mixed culture of AOB	Nonylphenol	–	93%	Wang et al. (2015)
5	*Nitrosomonas europaea*	Triclosan	–	100%	Roh et al. (2009)

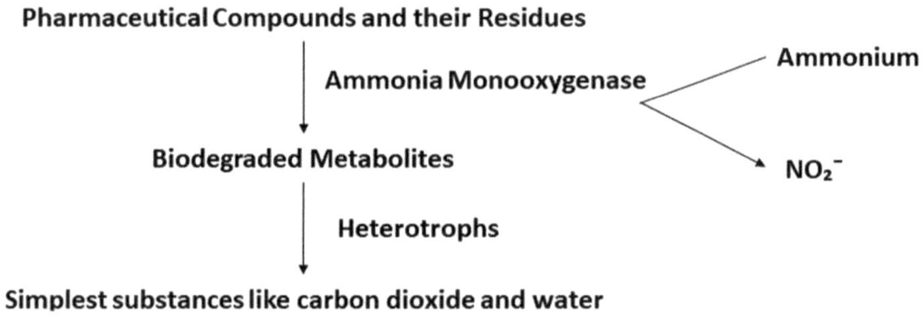

Fig. 1 Cometabolic biodegradation mechanism of AOB followed to breakdown pharmaceutical complexes

bioformulation for methanol production in industrial scale. Application of AOB might lead to the reduction of energy consumption in the wastewater treatment plants. Since methanol is widely used in increasing the denitrification activity in the wastewater treatment processes, thus, AOB bioformulation might play an effective role in the technique.

References

Anthonisen, A. C., Loehr, R. C., Prakasam, T. B., & Srinath, E. G. (1976). Inhibition of nitrification by ammonia and nitrous acid. *Journal (water Pollution Control Federation), 1*, 835–852.

Arp, D. J., Sayavedra-Soto, L. A., & Hommes, N. G. (2002). Molecular biology and biochemistry of ammonia oxidation by *Nitrosomonas europaea*. *Archives of Microbiology 178*(4), 250–255

Dawas-Massalha, A., Gur-Reznik, S., Lerman, S., Sabbah, I., & Dosoretz, C. G. (2014). Co-metabolic oxidation of pharmaceutical compounds by a nitrifying bacterial enrichment. *Bioresource Technology, 1;167*, 336–342

Fowler, D., Pyle, J. A., Raven, J. A., & Sutton, M. A. (2013). The global nitrogen cycle in the twenty-first century: Introduction. *Philosophical Transactions of the Royal Society b: Biological Sciences, 368*(1621), 20130165.

Gao, J. F., Luo, X., Wu, G. X., Li, T., & Peng, Y. Z. (2013). Quantitative analyses of the composition and abundance of ammonia-oxidizing archaea and ammonia-oxidizing bacteria in eight full-scale biological wastewater treatment plants. *Bioresource Technology, 1*(138), 285–296.

Ge, X., Yang, L., Sheets, J. P., Yu, Z., & Li, Y. (2014). Biological conversion of methane to liquid fuels: status and opportunities. *Biotechnology Advances 1;32*(8), 1460–1475

Gilch, S., Meyer, O., & Schmidt, I. (2010). Electron paramagnetic studies of the copper and iron containing soluble ammonia monooxygenase from *Nitrosomonas europaea. BioMetals, 23*(4), 613–622.

Hanaki, K., Wantawin, C., & Ohgaki, S. (1990). Nitrification at low levels of dissolved oxygen with and without organic loading in a suspended-growth reactor. *Water Research, 24*(3), 297–302.

Jones, R. D., & Morita, R. Y. (1983). Methane oxidation by Nitrosococcus oceanus and Nitrosomonas *europaea. Applied and Environmental Microbiology, 45*(2), 401–410.

Kassotaki, E., Buttiglieri, G., Ferrando-Climent, L., Rodriguez-Roda, I., & Pijuan, M. (2016). Enhanced sulfamethoxazole degradation through ammonia oxidizing bacteria co-metabolism and fate of transformation products. *Water Research. 1;94*, 111–119

Koops, H. P., & Pommerening-Röser, A. (2001). Distribution and ecophysiology of the nitrifying bacteria emphasizing cultured species. *FEMS Microbiology Ecology, 37*(1), 1–9.

Kumwimba, M. N., & Meng, F. (2019). Roles of ammonia-oxidizing bacteria in improving metabolism and cometabolism of trace organic chemicals in biological wastewater treatment processes: A review. *Science of the Total Environment, 1*(659), 419–441.

Limpiyakorn, T., Sonthiphand, P., Rongsayamanont, C., & Polprasert, C. (2011). Abundance of amoA genes of ammonia-oxidizing archaea and bacteria in activated sludge of full-scale wastewater treatment plants. *Bioresource Technology, 102*(4), 3694–3701.

Mohammed, R. N., Abu-Alhail, S., & Xi-Wu, L. (2014). Long-term operation of a novel pilot-scale six tanks alternately operating activated sludge process in treating domestic wastewater. *Environmental Technology 3;35*(15), 1874–1885

Purkhold, U., Pommerening-Röser, A., Juretschko, S., Schmid, M. C., Koops, H. P., & Wagner, M. (2000). Phylogeny of all recognized species of ammonia oxidizers based on comparative 16S rRNA and amoA sequence analysis: Implications for molecular diversity surveys. *Applied and Environmental Microbiology, 66*(12), 5368–5382.

Roh, H., Subramanya, N., Zhao, F., Yu, C. P., Sandt, J., & Chu, K. H. (2009). Biodegradation potential of wastewater micropollutants by ammonia-oxidizing bacteria. *Chemosphere, 77*(8), 1084–1089.

Shah, M. P. (2020). *Microbial bioremediation and biodegradation.* Springer.

Shah, M. P. (2021). *Removal of refractory pollutants from wastewater treatment plants.* CRC Press.

Shi, J., Fujisawa, S., Nakai, S., & Hosomi, M. (2004). Biodegradation of natural and synthetic estrogens by nitrifying activated sludge and ammonia-oxidizing bacterium Nitrosomonas *europaea. Water Research, 38*(9), 2323–2330.

Soliman, M., & Eldyasti, A. (2018). Ammonia-Oxidizing Bacteria (AOB): Opportunities and applications—a review. *Reviews in Environmental Science and Bio/technology, 17*(2), 285–321.

Su, Q., Schittich, A. R., Jensen, M. M., Ng, H., & Smets, B. F. (2021). Role of ammonia oxidation in organic micropollutant transformation during wastewater treatment: Insights from molecular, cellular, and community level observations. *Environmental Science and Technology, 55*(4), 2173–2188.

Sun, Q., Li, Y., Chou, P. H., Peng, P. Y., & Yu, C. P. (2012). Transformation of bisphenol A and alkylphenols by ammonia-oxidizing bacteria through nitration. *Environmental Science and Technology., 46*(8), 4442–4448.

Wang, X., Wen, X., Xia, Y., Hu, M., Zhao, F., & Ding, K. (2012). Ammonia oxidizing bacteria community dynamics in a pilot-scale wastewater treatment plant. *PLoS ONE, 7*(4), e36272.
Wang, Z., Zong, H., Zheng, H., Liu, G., Chen, L., Xing, B. (2015). Reduced nitrification and abundance of ammonia-oxidizing bacteria in acidic soil amended with biochar. *Chemosphere, 138*, 576–583.
Wei, D., Du, B., Xue, X., Dai, P., & Zhang, J. (2014). Analysis of factors affecting the performance of partial nitrification in a sequencing batch reactor. *Applied Microbiology and Biotechnology, 98*(4), 1863–1870.
Xu, X., Chen, X., Su, P., Fang, F., & Hu, B. (2016). Biodegradation potential of polycyclic aromatic hydrocarbons by bacteria strains enriched from Yangtze River sediments. *Environmental Technology., 37*(5), 513–520.
Zhang, Y., Chen, L., Sun, R., Dai, T., Tian, J., & Wen, D. (2015). Ammonia-oxidizing bacteria and archaea in wastewater treatment plant sludge and nearby coastal sediment in an industrial area in China. *Applied Microbiology and Biotechnology, 99*(10), 4495–4507.

Ammonia-Oxidizing Bacteria: Biochemical and Molecular Characteristics

Esra Şentürk, Gülsüm Atasoy, and Pınar Şanlıbaba

Abstract Ammonia-oxidizing bacteria (AOB) play a fundamental role in the biogeochemical cycle of nitrogen. AOB reduce the levels of ammonia present in the atmosphere through the nitrification process. Ammonia is released into the environment mainly due to the decomposition of organic matter or the application of NH_3-based fertilizers to agricultural fields. AOB oxidize this ammonia and convert it into nitrite, which serves as an energy source for the growth of plants and microorganisms, particularly in environments with low oxygen levels. AOB also contribute to maintaining environmental quality by assisting the decomposition process occurring in nature. Therefore, the presence of AOB is generally desirable in both natural and anthropogenic industrial systems as it provides a mechanism for regulating the loss of nitrogen from the environment. Moreover, AOB are ubiquitous in the environment; they are present in the soil, freshwater, and marine habitats, anthropogenically designed ecosystems such as wastewater treatment plants, and even on human skin. The functional significance of this group of bacteria renders it important to study their ecology and physiology, which have, consequently, been undertaken as subjects of intensive research in recent years. In this context, the present report aims to provide relevant information regarding the biochemical and molecular characteristics of ammonia-oxidizing bacteria.

Keywords Ammonia-oxidizing bacteria · Biochemical characteristics · Molecular characteristics · Growth parameters · Functional characteristics

1 Introduction

The nitrification reaction is a two-step process in which ammonia (NH_4) is oxidized into nitrate (NO_3^-) via nitrite (NO_2) (Kumwimba & Meng, 2019). The microbial oxidation of ammonia into nitrite by different groups of microorganisms is the first

E. Şentürk · G. Atasoy · P. Şanlıbaba (✉)
Department of Food Engineering, Ankara University, Engineering Faculty 50th Year Settlement, 06830 Gölbaşı, Ankara, Turkey
e-mail: sanlibab@ankara.edu.tr

M. P. Shah (ed.), *Anammox Technology in Industrial Wastewater Treatment*,
https://doi.org/10.1007/978-981-99-3459-1_2

and the rate-limiting step of the nitrification reaction and a critical part of the global biogeochemical nitrogen cycle (Chu et al., 2007; Lehtovirta-Morley, 2018; Wang et al., 2020).

The first step of the nitrification reaction (Eq. 1) comprises the conversion of ammonium into nitrite by the ammonia-oxidizing bacteria (AOB). The second step of this reaction (Eq. 2) is the conversion of nitrite into nitrate by the nitrite-oxidizing bacteria (NOB) (Soliman & Eldyasti, 2018).

$$NH_3^+ + 1,5O_2 \xrightarrow{AOB} NO_2^- + H^+ + H_2O \tag{1}$$

$$NO_2^- + 0,5O_2 \xrightarrow{NOB} NO_3^- \tag{2}$$

Although it is generally accepted that nitrification is a two-step process conducted by two phylogenetically different groups of microorganisms, this conventional view has recently been challenged after the discovery of ammonia-oxidizing archaea by Könneke in 2005 and the subsequent discovery of the bacteria capable of oxidizing ammonia directly to nitrate in 2015 (Osburn & Barrett, 2020).

In 1877, Schloesing and Muntz demonstrated that soil-based nitrification was inhibited by the application of chloroform. These authors reported the first evidence for the oxidation of ammonia being a biological process. Subsequently, in 1890, Frankland and Frankland pioneered culturing the ammonia-oxidizing bacterial species *Nitrosomonas europaea*, which is considered the representative of the ammonia-oxidizing bacteria even to this day; the cultured bacterial strain had been isolated by Sergei Winogradsky. Since then, the ammonia-oxidizing bacteria were assumed to be the sole drivers of the nitrification reaction until the first archaeon capable of oxidizing ammonia was reported in 2005 (Lehtovirta-Morley, 2018).

Thereafter, it has been generally accepted that the ammonia-oxidizing archaea (AOA) and AOB were jointly responsible for the oxidation of ammonia (Wang et al., 2020). However, recent studies have revealed three different groups of aerobic autotrophic microorganisms that are capable of oxidizing ammonia: *(i)* AOB, *(ii)* AOA, and *(iii)* complete ammonia oxidizer (Comammox) bacteria. The Comammox group is reported to enable the complete oxidation of ammonia into nitrate (Fig. 1) (Lehtovirta-Morley, 2018).

Ammonium is a common pollutant in both natural and anthropological environments (Bai et al., 2012). N_2O, a greenhouse gas, is reported to be approximately

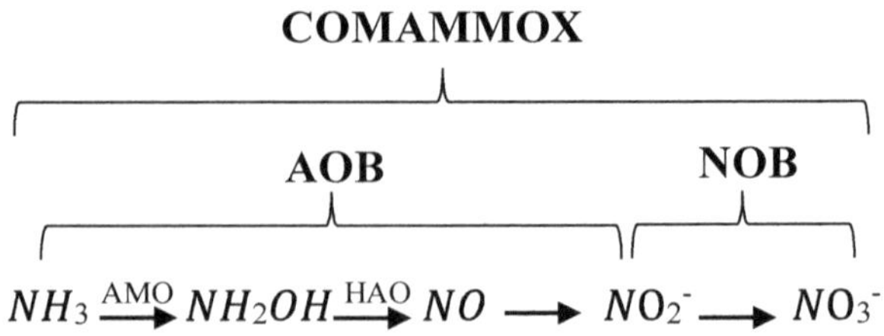

Fig. 1 The microorganisms responsible for the nitrification reaction (Stein, 2019) (AMO: Ammonia monooxygenase; HAO: Hydroxylamine oxidoreductase)

300 times more effective in destroying the ozone layer compared to carbon dioxide (Kim et al., 2010). Nitrification is the biological oxidation of ammonium into nitrite and nitrate and forms an important part of the environmental nitrogen cycle and the wastewater treatment processes (Zhang et al., 2011; Shah, 2020, 2021). Therefore, it is important to study the biological oxidation of ammonium and the microorganisms that are responsible for it. In this context, the present report summarizes the general and molecular characteristics of ammonia-oxidizing bacteria, an important class of microorganisms responsible for conducting the biological oxidation of ammonia.

2 General Characteristics of Ammonia-Oxidizing Bacteria

AOB are important members of the terrestrial, marine, and industrial microbial communities of bacteria that are involved in the first and the rate-limiting step of the nitrification reaction. AOB, therefore, play a fundamental role in the nitrogen cycle occurring in these systems (Dong et al., 2014; Zorz et al., 2018). AOB are chemolithoautotrophs, i.e., they use ammonia as an electron donor and carbon dioxide as the sole carbon source for respiration and cell synthesis (Guo et al., 2013; Keluskar et al., 2013). AOB are autotrophic bacteria that fix carbon through the Calvin-Benson-Bassham (CBB) cycle (Sedlacek et al., 2016; Zorz et al., 2018). Certain types of AOB are tolerant to both aerobic and anaerobic conditions, although the energy required for their growth is generally obtained from aerobic oxidation (Kumwimba & Meng, 2019).

2.1 *Ecology of Ammonia-Oxidizing Bacteria*

AOB, which were previously considered the only group of microorganisms responsible for the autotrophic oxidation of ammonia for a long time, occur in a wide range of environments, demonstrating a wide distribution in both marine and terrestrial habitats (Kumwimba & Meng, 2019). AOB have been detected in the soil, freshwaters, salt waters, wastewater treatment plants, and even on human skin. The AOB belonging to the β-Proteobacterial genus are generally present in soil environments and wastewater treatment plants, while the γ-Proteobacterial AOB are reported mainly from the marine habitats (Lehtovirta-Morley, 2018). Molecular studies have revealed that AOB exist in habitats with pH values of ~4, such as the Antarctic soils, acidic tea soil, and forest soils, in the environments with a pH value of 10, such as soda lakes, and also in extreme environments, such as terrestrial hot springs and hypersaline habitats. The AOB species have also been detected in various artificial environments, including wastewater treatment plants (WWTPs), wetlands, aquaculture biofiltration units, drinking water treatment plants, etc. Molecular biology studies have revealed a dominance of *Nitrosomonas europaea* and *Nitrosomonas eutropha* in the hypersaline ecosystems. The species of *Nitrosospira* and *Nitrosomonas* reportedly dominate the alkaline and neutral soils, while *Nitrosomonas oligotropha*, *Nitrosomonas europaea*, and *Nitrosomonas nitrosa* are the most dominant ones in different domestic wastewater treatment facilities (Kumwimba & Meng, 2019). Terrestrial AOB are generally limited to the genus β-*Proteobacteria*, while marine AOB include species from both *β-Proteobacteria* and γ-*Proteobacteria*. The AOB that usually dominate the soils are those from the *Nitrosospira* spp. In marine habitats, species of *Nitrosococcus, Nitrosomonas*, and *Nitrosospira* coexist with the ammonia-oxidizing *Archaea* (Norton, 2011).

2.2 *Parameters Affecting the Growth of Ammonia-Oxidizing Bacteria*

An effective nitrification reaction is achieved by ensuring the development of AOB and inhibition of NOB. The successful growth of AOB relies on the understanding of the various parameters influencing their development. Studies have demonstrated that the activity, distribution, and community structure of AOB are sensitive to different environmental conditions, including the ammonia concentration, pH, dissolved oxygen (DO) concentration, temperature, salinity, and inhibitors (Guo et al., 2013; Soliman & Eldyasti, 2018; Stein, 2019). Environmental pH and ammonia concentration are the two main factors affecting the diversity and abundance of ammonia oxidizers. The additional factors that affect the environmental distribution of ammonia-oxidizing microorganisms include organic compounds, metals, temperature, and salinity (Lehtovirta-Morley, 2018). In addition, the wastewater characteristics, reactor types, and process operating systems impact the abundance and the community structure of AOB (Guo et al., 2013).

2.2.1 Ammonia Concentration

AOB and AOA are often present in the same environment. The difference is that AOB prefer environments with a high substrate while AOA prefer environments with a lower substrate. The rate of development of AOB increases with an increase in the concentrations of ammonia and oxygen. Increasing the concentration of ammonia may lead to a direct increase in the total number of AOB and also in the ammonia-oxidation activity of AOB. However, excessive free ammonia (FA) may inhibit the growth of AOB (Guo et al., 2013). It is recognized that AOB prefer a high concentration of ammonia in the soil while AOA predominate when the ammonia concentration is low. AOB are, therefore, predominant in the wastewater treatment plants, which have a high ammonium concentration. In marine ecosystems, ammonia concentrations are generally very low. It is reported that the AOB *Nitrosomonas europaea* is capable of using only ammonia and not ammonium, and is, therefore, unable to thrive in acidic environments (Lehtovirta-Morley, 2018).

2.2.2 Dissolved Oxygen (DO) Concentration

Since the process of oxidation of ammonia occurs under aerobic conditions, decreasing the concentration of dissolved oxygen decreases the rate of ammonia oxidation. It is reported that certain species that were isolated from oxygen-enriched habitats exhibited no ammonia-oxidizing activity under anaerobic conditions (Guo et al., 2013). Therefore, controlling the dissolved oxygen concentration in a reactor is a possible approach to increase nitrite deposition and, consequently, accumulate AOB (Soliman & Eldyasti, 2018).

2.2.3 Temperature

Temperature exerts a great influence on the oxidation of ammonia in the nitrification reaction. It is reported that nitrification proceeds better in warmer seasons (Soliman & Eldyasti, 2018). Temperature has a direct impact on bacterial activity, free ammonia concentration, pH, and moisture, each of which may indirectly influence the abundance and the community structure of AOB. Although the activity and the development of AOB are influenced by the temperature of the environment in which these bacteria are dwelling, AOB are capable of adapting to a wide temperature range of 5–40 °C. However, the rate of ammonia oxidation declines rapidly at temperatures below 15 °C or those above 40 °C (Guo et al., 2013).

2.2.4 PH

The pH of the environment affects the balance of free ammonia (FA) and free nitrous acid (FNA), which, in turn, has an inhibitory effect on both AOB and NOB. Therefore, FA and FNA are commonly used for regulating pH to achieve partial nitrification. When a higher free ammonia concentration is required for AOB inhibition, it indicates that the nitrite oxidizers are more sensitive to free ammonia (Soliman & Eldyasti, 2018). The activity of the cytoplasmic enzyme of AOB and the form in which ammonia exists in the system may be affected at an ambient pH. *Nitrosomonas* species exhibit optimum growth in the pH range of 6.0–8.0, while the members of the *Nitrosospira* and *Nitrosococcus* genera predominate under low pH conditions (Guo et al., 2013).

2.2.5 Salinity and Inhibitors

Salt is necessary for the growth of AOB. However, high salinity leads to growth inhibition in these bacteria. The other substances that inhibit the growth of AOB include heavy metals, organic compounds, toxins, fulvic acids, volatile fatty acids, and halides (Guo et al., 2013).

3 Ammonia-Oxidizing Bacteria Diversity

3.1 Aerobic Ammonia-Oxidizing Bacteria

Since the first isolation of AOB in 1890, numerous studies have been conducted to determine the phylogenetic diversity of these bacteria (Guo et al., 2013; Kozlowski et al., 2016a; Soliman & Eldyasti, 2018). While the same basic metabolic processes occur in all aerobic AOB, large ecophysiological differences exist among different strains (Bernhard & Bollmann, 2010).

All the recognized AOB capable of oxidizing ammonia to nitrite are grouped into the major categories of *β-Proteobacteria* and *γ-Proteobacteria* in the *Proteobacteria* phylum. The *γ-Proteobacteria* AOB belong to the *Chromatiaceae* family and include *Nitrosococcus* and *Nitrosolobus*. While *Nitrosococcus* is a genus of halophilic bacteria that occur in marine habitats and salt lakes, the genus *Nitrosolobus* comprises the acid-tolerant AOB present in acidic soils. In contrast, the AOB that generally predominate in the terrestrial and freshwater ecosystems are classified in the *Nitrosomonadaceae* family belonging to *β-Proteobacteria* (Lukumbuzya et al., 2020; Schebor, 2014).

A total of five AOB genera have been identified so far and classified into the *Proteobacteria* class (Fig. 2).

Nitrosomonas (including *Nitrosococcus mobilis*), *Nitrosospira, Nitrosovibrio*, and *Nitrosolobus* are classified into the *β-Proteobacteria* group, while *Nitrosococcus* belongs to the *γ-Proteobacteria* subclass (Díaz et al., 2019; Guo et al., 2013; Kumwimba & Meng, 2019; Soliman & Eldyasti, 2018). The genus *Nitrosospira* was modified to include the genera *Nitrosovbrio* and *Nitrosolobus* based on the 16S rRNA gene sequences. Therefore, Beta-proteobacterial AOB can be reclassified to consist of the genera *Nitrosomonas* and *Nitrosospira* (Campbell et al., 2011; Fujitani et al., 2015; Hayatsu et al., 2017; Holmes et al., 2019; Lukumbuzya et al., 2020; Purkhold et al., 2000; Rice et al., 2016; Schebor, 2014; Sedlacek et al., 2016, 2019).

The cells in AOB may be classified into four main categories based on their morphological characteristics, such as cell shape, size, motility, and intracytoplasmic membranes. The first category is the *Nitrosomonas* clusters, which include 0.7–1.5 μm × 1.0–2.4 μm in size, polar to subpolar motile, straight rod cells with peripheral flattened vesicles. The second category is the *Nitrosococcus* species, which have spherical ellipsoidal cells, 1.5–1.8 μm × 1.7–2.5 μm in size, with a tuft of flagella for motility and peripheral or central vesicle clumps. The third category is the *Nitrosospira* species, in which peritric motile cells of size 0.3–0.8 μm × 1.0–8.0 μm are tightly coiled in spirals, with invaginations in intracytoplasmic membranes. The fourth category is the *Nitrosolobus* species, which have pleomorphic lobate cells of size 1.0–1.5 μm × 1.0–2.5 μm, with peritric flagella and a segmented intracytoplasmic membrane. In addition, *Nitrosovibrio* species is reported, which has polar or subpolar arcuate rods of size 0.3–0.4 μm × 1.1–3.0 μm, with invaginations in the intracytoplasmic membrane (Soliman & Eldyasti, 2018).

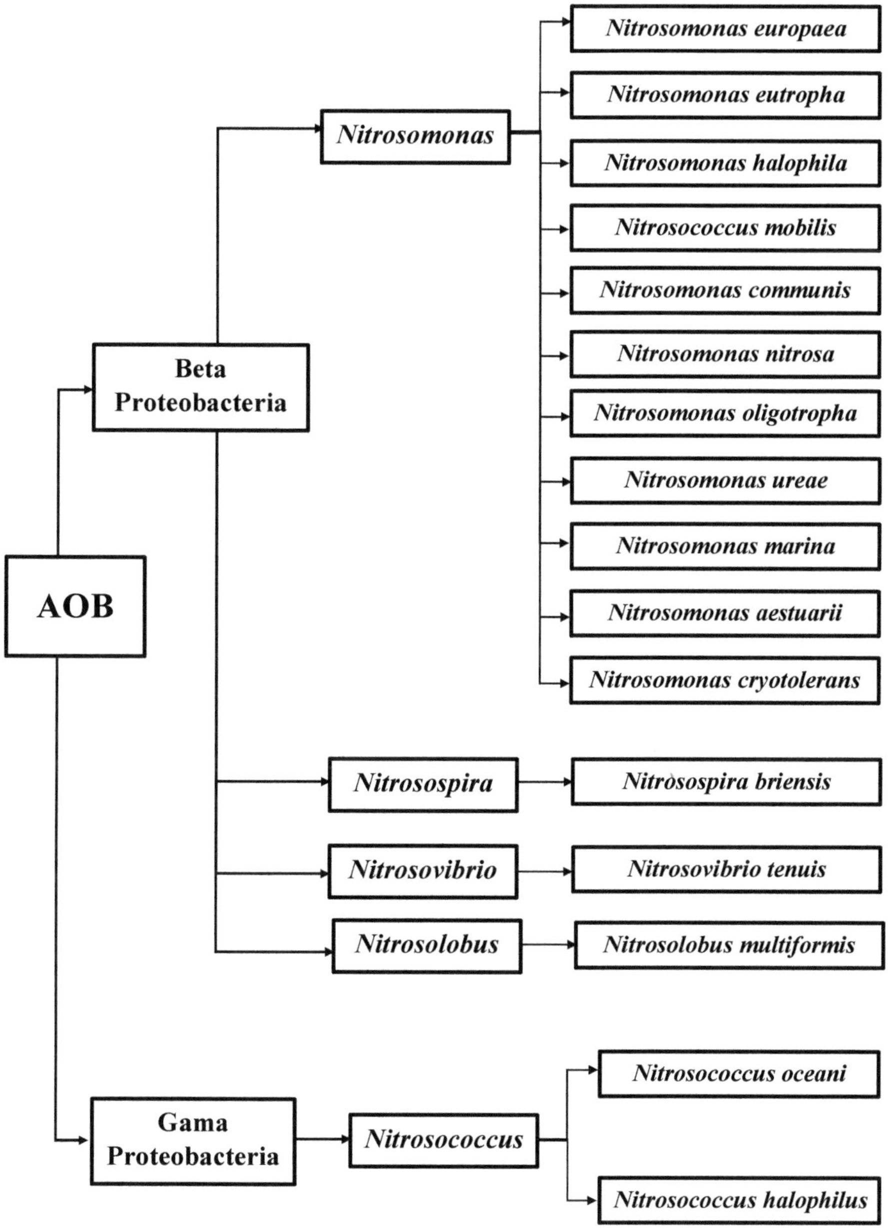

Fig. 2 Types of AOB (Image adapted from: Guo et al., 2013; Soliman & Eldyasti, 2018)

AOB may broadly be categorized into three groups, each with its unique ecophysiological parameters and preferred habitats. The first group is the genus *Nitrosomonas*, which comprises six lineages. The first *Nitrosomonas* lineage includes four species (*Nitrosomonas europaea, Nitrosomonas eutropha, Nitrosomonas halophila,* and *Nitrosococcus mobilis*) and is characterized by moderate salt requirement and negative urease activity. This lineage has a substrate affinity of 30–61 μM and is usually isolated from sewage disposal facilities, eutrophic freshwaters, and salt waters. The second lineage of *Nitrosomonas* includes three species (*Nitrosomonas communis*, *Nitrosomonas* sp. I, and *Nitrosomonas* sp. II) and does not require salt. It also exhibits a lower substrate affinity of 14–43 μM and has no urease activity. It mainly occurs in non-acidic soils. The third lineage of *Nitrosomonas* includes the *Nitrosomonas nitrosa* species, which has no salt requirement and exhibits a positive urease activity. It exhibits a moderate substrate affinity of 19–46 μM and usually occurs in eutrophic freshwaters. The fourth lineage of *Nitrosomonas* includes two species (*Nitrosomonas ureae* and *Nitrosomonas oligotropha*) and has no salt requirement. This lineage exhibits the lowest substrate affinity of 1.9–4.2 μM and positive urease activity. Its preferred habitats are oligotrophic freshwaters and natural salt waters. Finally, the strains of *Nitrosomonas marina*, *Nitrosomonas* sp. III, *Nitrosomonas aestuarii*, and *Nitrosomonas cryotolerans* constitute the fifth and the sixth linages of *Nitrosomonas*, which exhibit the highest substrate affinities of 50–52 and 42–59 μM, respectively. These bacteria are obligately halophilic and exhibit a positive urease activity. These are commonly present in marine environments. The second group of AOB, which includes *Nitrosolobus multiformis*, *Nitrosovibrio tenuis*, and *Nitrosospira* sp. I, do not require salt, exhibit similar ecophysiological parameters, and occur in soil, rocks, and freshwaters. The third group of AOB comprises the *Nitrosococcus oceani* species, which is an obligate halophile, exhibits a positive urease activity, and occurs in marine environments, and *Nitrosococcus halophilus*, which exhibits a negative urease activity (Soliman & Eldyasti, 2018).

Nitrosomonas and *Nitrosospira*, which comprise phylogenetically distinct clusters with unique physiological characteristics, are the most widely distributed AOB in nature. The physiological, biochemical, and genetic properties of pure strains isolated from each cluster have been investigated by several researchers. Accordingly, Cluster 6A is represented by the *Nitrosomonas oligotropha* lineage, which comprises members that are relatively ammonia-sensitive and present in freshwaters, wastewaters, and terrestrial systems. The members of Cluster 6A are characterized by their high substrate affinity and sensitivity to high concentrations of ammonium and nitrite. Cluster 6B, on the other hand, includes *Nitrosomonas marina* and *Nitrosomonas aestuarii*, which exhibit high salt tolerance and are, therefore, often isolated from marine systems. Cluster 7 includes *Nitrosomonas europaea*, *Nitrosomonas eutropha*, and other pure strains capable of tolerating high ammonia concentrations. The members of this cluster, which may be isolated from various environments, are characterized by low substrate affinity and tolerance to high concentrations of both ammonium and nitrite. *Nitrosomonas mobilis*, an important AOB species dominant in various wastewater treatment systems, also belongs to this cluster. Several pure strains of this species are reportedly isolated from nitrification biofilms (Fujitani et al., 2015; Sedlacek et al., 2019). *Nitrosomonas europaea* and *Nitrosomonas mobilis* (cluster 7) are predominant in the wastewater treatment bioreactors loaded with high concentrations of ammonia and nitrite, while *Nitrosomonas oligotropha* (cluster 6A) dominates the systems with a lower ammonia concentration (Thandar et al., 2016).

Nitrosomonas europaea requires only NH_3, CO_2, and mineral salts for its growth and is, therefore, an obligate chemolithoautotroph. *Nitrosomonas europaea* is a member of the *β-Proteobacteria*, a large group of eubacteria assumed to be of photosynthetic origin. *Nitrosomonas europaea* is the best-studied among all the ammonia-oxidizing bacteria at the molecular level. First isolated by Winogradsky in 1892, *Nitrosomonas europaea* occurs in a variety of natural and designed habitats, including soil, freshwaters, sediment, estuary, and activated sludge (Park & Noguera, 2007). The genome sequence of *Nitrosomonas europaea* is 2.8 Mb long and contains nearly 2700 open reading frames (ORF). *Nitrosomonas europaea* is a motile microorganism capable of forming biofilms (Schmidt et al., 2004). Both *Nitrosomonas europaea* and *Nitrosomonas eutropha* are halotolerant and capable of surviving in NaCl up to a concentration of 400 mM. Both the species lack urease activity. *Nitrosomonas europaea* ATCC 19,718 may be enhanced through chemolithoheterotrophy using ammonia as the energy source and fructose as the carbon source (Stein et al., 2007). The *Nitrosomonas europaea* species cluster of AOB presents the greatest physiological and phylogenetic diversity among all kinds of AOB. Certain other species of AOB, such as *Nitrosomonas marina*, *Nitrosomonas aestuarii*, and *Nitrosomonas cryotolerans*, are obligate halophiles (Bernhard & Bollmann, 2010).

Cluster 8 (the *Nitrosomonas communis* lineage) AOB are mainly isolated from eutrophic soil and aquatic environments. Cluster 9 AOB include the *Nitrosomonas* sp. Nm143 and *Nitrosomonas cyrotolerans* species. These are commonly present in marine or other saline environments (Sedlacek et al., 2019). *Nitrosomonas communis* Nm2 occurs commonly in terrestrial ecosystems and was first isolated in Corfu, Greece. This is a mesophilic aerobic β-*Proteobacterium* species that oxidizes ammonia to nitrite as a sole source of energy and reducing agent. Although the genomes of certain *Nitrosospira* contain complete urease genes, *Nitrosomonas communis* encodes only urea carboxylase and allophanate hydrolase. While the genomes of a majority of the AOB contain the copper-containing nitrite reductase gene *nirK*, no homologs of the nitrite reductases *nirK* and *nirS* are reported in *Nitrosomonas communis* (Kozlowski et al., 2016c).

According to the observed complexity of the AmoA and 16S rRNA genes, most AOB are classified into genus *Nitrosomonas*, with *Nitrosomonas ureae*, *Nitrosomonas oligotropha*, *Nitrosomonas marina*, and *Nitrosomonas aestuarii* being the four dominant species (Zhang et al., 2011). Under certain conditions, such as low pH, certain AOB may utilize urea as both nitrogen and carbon source. *Nitrosomonas ureae* is one of the AOB species that is capable of utilizing urea (Zorz et al., 2018) and may utilize urea as an alternative nitrogen source. This species contains genes for both urea carboxylase and an allophanate hydrolase, as well as genes for a complete urease present in certain *Nitrosospira* genomes. *Nitrosomonas ureae* Nm10, an oligotrophic aerobic β-*Proteobacterium* belonging to the *Nitrosomonas* Cluster 6a, was first isolated from soil specimens in Italy. Its genome is 3.3 Mbp in length, with an average G + C content of 44.5% and 2897 estimated protein-coding genes (Kozlowski et al., 2016b).

The γ-*Proteobacterium* species *Nitrosococcus oceani* was the first obligate ammonia-oxidizer isolated from seawater (Stein et al., 2013). Its genome comprises a single circular chromosome with 3,481,691 bp, a G + C content of 50.4%, and 3052 estimated protein-coding genes, along with a plasmid with 40,420 bp and 41 estimated protein-coding genes. *Nitrosococcus oceani* ATCC 19,707, a member of Chromatiales, also known as the purple sulfur bacteria, is the first AOB isolated from seawater using the enrichment culture technique. Besides seawater, *Nitrosococcus oceani* has been detected in the brine of Lake Bonney, a permanently ice-covered lake in Antarctica, using the immunofluorescence and fluorescence in-situ hybridization methods (Klotz et al., 2006).

The complete genome of the ammonia-oxidizing bacterial species *Nitrosospira multiformis* ATCC 25,196 contains a circular chromosome and three small plasmids, together containing 3,234,309 bp and 2827 protein-encoding genes. The characteristics that distinguish *Nitrosospira multiformis* from *Nitrosomonas europaea* include the presence of gene clusters encoding urease and hydrogenase (a ribulose-bisphosphate carboxylase/oxygenase-encoding operon with a distinctive structure and phylogeny) and a relatively small complement of genes associated with Fe acquisition. *Nitrosospira multiformis* and other *Nitrosospira* species are recognized as important members of the AOB community in agricultural soils of various geographical locations (Norton et al., 2008). The 16S rRNA gene phylogeny analysis revealed that the genus *Nitrosospira* is represented by five clusters, with *Nitrosospira briensis*, *Nitrosospira tenuis*, and *Nitrosospira multiformis* belonging to cluster 3. The only publicly available *Nitrosospira* genome is that of *Nitrosospira multiformis* ATCC 25,196 (Garcia et al., 2013).

Numerous AOB isolated from or detected in aerobic agricultural surface soils using non-culture methods are members of the genus *Nitrosospira*. *Nitrosospira briensis* C-128 is a chemolithoautotrophic ammonia-oxidizing Betaproteobacteria that was isolated from an acidic agricultural soil for growing blueberry in the USA (Rice et al., 2016).

3.2 *Anaerobic Ammonia-Oxidizing Bacteria*

Anaerobic ammonia oxidation (anammox) was discovered in the early 1990s in a denitrification fluidized bed reactor in the Netherlands. Anammox involves the oxidation of ammonium to the dinitrogen gas using nitrite as the electron acceptor under anoxic conditions (Ding et al., 2013).

The stoichiometry of the Anammox processes was first established in 1998 based on the results of the laboratory-scale chemostat experiments that involved catalyzing ammonium, nitrite, and bicarbonate. The literature reports seven known species of anammox bacteria, which are classified into phylum Planctomycetes and class Brocadiales—Brocadia, Kuenenia, Anammoxoglobus, Jettenia, Scalindua, Anammoxomicrobium, and Brasilis. Most of these genera originally discovered in different habitats have been detected in the wastewater treatment systems. The first anammox bacterial species discovered was *Brocadia anammoxidans* (Sarı & Mertoğlu, 2019), which comprises coccoid bacteria less than 1 μm in diameter. These bacteria reproduce through budding and have a growth period of 10–30 days (Esen, 2013).

The activity and distribution of Anammox are affected by different environmental conditions, including salinity, temperature, and the concentrations of DO, ammonia, nitrite, organic carbon, and nitrate. While the anammox process usually occurs in mesophilic environments at temperatures ranging from 6 °C to 43 °C, the anammox activity decreases rapidly at temperatures below 15 °C or those above 40 °C. However, increasing evidence suggests that the anammox bacteria are capable of tolerating a relatively wider temperature range of –2 °C to 85 °C (Guo et al., 2013). Similar to the temperature parameter, the pH range in which the anammox bacteria are active is also wide, with the optimum activity exhibited in the pH range of 6.7–8.3 (Sarı & Mertoğlu, 2019). A study involving the adaptation of the anammox population to high-salinity wastewater demonstrated that these bacteria also had a good degree of tolerance to salinity (Guo et al., 2013).

While the Anammox bacteria that conduct the anammox reaction are distributed in different parts of the natural anoxic ecosystems, such as marine sediments and water columns, these are also common in artificial ecosystems such as wastewater treatment plants (Ding et al., 2013). The Anammox bacteria have been detected in soils, although the relevant information available currently regarding their distribution, diversity, and activity in the terrestrial systems is relatively limited. Moreover, with the increase in the human knowledge of the different habitats or ecological niches of Anammox bacteria, these bacteria are being discovered in various environments, including marine, coastal, and estuarine sediments (Guo et al., 2013).

Currently, several suitable molecular biology methods are available for the detection of Anammox bacteria and the determination of their activity and functions in various ecosystems. PCR amplification using a specific Anammox primer or a specific Anammox reverse primer is employed commonly to detect the Anammox bacteria in the environment. The genes encoding key enzymes, such as hzo and nirS, which serve as functional biomarkers, are proposed as effective tools for detecting the Anammox bacteria and determining their diversity (Guo et al., 2013). Anammox bacteria contain a homolog of HAO, which, similar to AOB, oxidizes hydroxylamine to NO (Lehtovirta-Morley, 2018). FISH, FISH-MAR, and Raman-FISH techniques are commonly employed for the quantification of Anammox bacteria or for analyzing their activity and growth at the level of a single cell (Guo et al., 2013).

Several studies have reported that the Anammox process contributes significantly to the production of nitrogen gas in certain marine environments and is, therefore, crucial for the global nitrogen cycle. Since the Anammox bacterial community plays a key role in the anammox process, further investigation aiming at a comprehensive understanding of the structure, function, and microbial interactions of these bacteria is warranted (Ding et al., 2013).

3.3 Complete Ammonia-Oxidizing Bacteria

Complete ammonia oxidizers (Comammox) were discovered first in 2015 as the microorganisms capable of performing the entire nitrification process alone, without requiring assistance or contribution from other microorganisms. Comammox completely oxidize ammonia and convert it into nitrate directly. The first Comammox to be described was a *Nitrospira* that was enriched through a trickling filter connected to an aquaculture system and a moderately thermophilic biofilm submerged in hot groundwater in a 1200 m deep oil well (Daims et al., 2016). Since then, Comammox have been discovered in a variety of habitats, including water, soil, and sediment. Comammox are classified as the members of genus Nitrospira, which were previously thought to be capable of conducting nitrite oxidation only. Recent evidence suggests that the Comammox *Nitrospira* contributes to the nitrification process occurring in the soil, although its abundance and functional significance in soil ecosystems are not completely understood so far (Osburn & Barrett, 2020).

The discovery of Comammox revealed that the metabolic division of labor is not essential in the nitrification process. This revelation would have far-reaching implications that would influence future studies on the microbiology of the nitrogen cycle (Daims et al., 2015). The Comammox bacteria reported to date include Candidatus *Nitrospira inopinata*, Candidatus *Nitrospira nitrosa*, Candidatus *Nitrospira nitrificans*, and *Nitrospirae* sp., all of which are classified phylogenetically into the *Nitrospira* genus, which is the most diverse functional group of NOB. Among the six currently recognized sub-lineages of *Nitrospira*, lineage II appears to be the most widely distributed group of NOB and contains all the Comammox organisms reported to date (Wang et al., 2017).

Comammox bacteria carry genes associated with the oxidation of both ammonia and nitrite (Wang et al., 2017). Although Comammox utilize the enzymes ammonia monooxygenase (AMO) and hydroxylamine dehydrogenase (HAO) for the oxidation of ammonia, these enzymes are different from their homologs in AOB (Daims et al., 2016).

While all the AOB reported so far have large intracytoplasmic membrane structures, Comammox lack such membrane structures similar to AOA and contain only a plasma membrane. Both AOB and Comammox oxidize hydroxylamine using hydroxylamine dehydrogenase. Studies have demonstrated that the Comammox Nitrospira is capable of adapting to low concentrations of ammonia, similar to AOA. Studies have also demonstrated that the Comammox *Nitrospira* may occur in soils with neutral and acidic pH, aquatic environments, and designed ecosystems, although further studies are required on the contribution of these bacteria to the nitrogen cycle and their environmental distribution and physiology (Lehtovirta-Morley, 2018).

4 Mechanism Underlying the Oxidation of Ammonia by Ammonia-Oxidizing Microorganisms

The first step of the nitrification process that is conducted by the AOB is referred to as the nitritation reaction. Nitritation involves the conversion of ammonium to nitrite and comprises two major steps. The first step (Eq. 3) of the nitritation reaction is the oxidation of ammonium (NH_4^+) to hydroxylamine (NH_2OH), a reaction catalyzed by membrane-bound AMO. In the second step (Eq. 4) of the nitritation reaction, NH_2OH is oxidized to nitrite (NO_2^-) under the catalysis of the HAO enzyme (Guo et al., 2013; Matsuno et al., 2013; Soliman & Eldyasti, 2018).

$$NH_3 + O_2 + 2H^+ + 2e^- \underrightarrow{AMO} NH_2OH + H_2O \tag{3}$$

$$NH_2OH + H_2O \underrightarrow{HAO} NO_2^- + 5H^+ + 4e^- \tag{4}$$

$$2H^+ + 0.5O_2 + 2e^- \rightarrow H_2O$$

4.1 *Main Enzymes of Ammonia-Oxidizing Bacteria*

The ability to utilize ammonia aerobically as the sole source of energy and reducing agent requires four specific proteins—AMO, HAO, cytochrome c554, and cytochrome c_M552 (Arp et al., 2007). AMO and HAO are the two key enzymes required during the oxidation of ammonia. These two are interdependent as both produce the required substrate and electrons (Guo et al., 2013).

In both AOB and AOA, the first step in the oxidation of ammonia occurs under the catalysis of AMO, a member of the copper membrane monooxygenase (CuMMO) family. It is reported that HAO co-occurs with AMO in all the lineages of AOB and that the phylogenetic relationships among the HAO proteins reflect those among the AMO proteins (Ward et al., 2021).

Bacteria-mediated oxidation of ammonia was initially assumed to be a two-step process, in which ammonia is first oxidized by AMO to hydroxylamine, which is then acted upon by HAO to produce nitrite. However, this view was challenged by the recent discovery of NO, not nitrite, being the reaction product of bacterial HAO (Caranto & Lancaster, 2017). The discovery of NO as an essential intermediate in AOB implies that the ammonia-oxidizing bacteria utilize a three-step ammonia oxidation pathway, which possibly involves a third, so far unidentified, enzyme that catalyzes the final step of the conversion of NO into nitrite. Caranto and Lancaster (2017) suggested that this missing enzyme could be NirK, a nitrite reductase involved in the denitrification of nitrite to NO, although this hypothesis has not been investigated experimentally so far (Lehtovirta-Morley, 2018).

4.1.1 Ammonia Monooxygenase

All the AOB reported so far have extensive intracytoplasmic membrane structures, while both AOA and Comammox lack such membrane structures and have a plasma membrane only. In the ammonia oxidation process, ammonia is oxidized to hydroxylamine by AMO, which is a membrane-bound enzyme belonging to a superfamily of ammonia-, methane-, and alkane-monooxygenases. AMO is fundamentally essential to ammonia oxidation and is the only enzyme in the ammonia oxidation pathway that is shared by all three major groups of ammonia-oxidizing microorganisms. The *amoA* gene (which encodes the A subunit of AMO) is the most widely used functional marker gene for ammonia oxidation, which has been quite beneficial in understanding the ecology of ammonia oxidizers. Since AMO, the key enzyme of the ammonia oxidation process is membrane-bound, the levels of membrane proteins are crucial for ammonia oxidation (Lehtovirta-Morley, 2018).

The copper protein AMO initiates the catabolism of ammonia by oxidizing ammonia to hydroxylamine (Arp et al., 2007). The AMO gene encodes the first enzyme in the oxidation of ammonia to nitrite via hydroxylamine. Three amoCAB operons have been detected in the genome of *Nitrosomonas* sp. Is79 (Bollmann et al., 2013). In *Bacteria* and *Archaea*, the amoA gene encodes the subunits of AMO (alpha-A) (Gao et al., 2018; Mukhtar et al., 2017). Since the ammonia monooxygenase subunit A gene (amoA) is highly conserved among autotrophic AOB, compared to the *amoB* and *amoC* genes, it is, therefore, frequently used as an indicator of the presence of autotrophic AOB (Matsuno et al., 2013).

AOB are obligate chemolithotrophs that derive energy and reducing power from the oxidation of ammonia, which begins with the conversion of ammonia to hydroxylamine. This first step is catalyzed by the enzyme AMO, which is specific to ammonia oxidizers. The genes encoding AMO have recently been used as alternative targets in the molecular analysis of AOB. The AMO enzyme comprises three subunits containing the active site of AMO: AmoA, which is a membrane-bound protein of size 27–30 kDa; AmoB, which has a size of 38–43 kDa, and AmoC, which has a size of 31.516 kDa. The *amoA* and *amoB* genes are always contiguous and preceded by a third gene, *amoC*. While all β-Proteobacterial AOB examined so far appear to contain up to three almost-identical copies of the amo operons, only one copy is reported in the γ-Proteobacterial ammonia oxidizers *Nitrosococcus oceani* and *Nitrosococcus* sp. C-113 (Calvo & Garcia-Gil, 2004).

4.1.2 Hydroxylamine Oxidoreductase

HAO is the second enzyme to act in the oxidation of ammonia. HAO catalyzes the oxidation of hydroxylamine to nitrite (Bollmann et al., 2013).

Hydroxylamine and nitric oxide (NO) are the two toxic intermediates in both bacterial and archaeal ammonia oxidation. In bacteria, periplasmic hydroxylamine dehydrogenase is the next enzyme to act after AMO. Both AOB and Comammox oxidize hydroxylamine using hydroxylamine dehydrogenase [formerly referred to as hydroxylamine oxidoreductase (HAO)] (Lehtovirta-Morley, 2018). HAO is also highly expressed, although at a level lower than that of AMO (Zorz et al., 2018). HAO is a water-soluble periplasmic protein that has only one subunit with a mass of 63 kDa, although it has a highly complex structure (Guo et al., 2013).

4.1.3 Other Enzymes

Nitrite reductase (NirK) and nitric oxide reductase (NorB) have been thought to play an important role in the production of nitrous oxide (N_2O) under the action of ammonia-oxidizing bacteria. Periplasmic copper-containing nitrite reductase (NirK) and a membrane-bound nitric oxide reductase (NorB) have been detected in the genome sequences of various AOB. However, significant gaps remain in deciphering the activity and functionality of NirK and NorB, due to which their precise role in NO production under the action of ammonia oxidizers remains unclear so far (Kozlowski et al., 2014).

Nitric oxide reductase is an important enzyme in the AOB nitrification–denitrification pathway. This enzyme converts NO to nitrous oxide (N_2O), which has a higher potency as a greenhouse gas compared to carbon dioxide. Homologs of nitric oxide reductase, namely, NorB and NorY, are present in all AOB, with *Nitrosomonas ureae* and *Nitrosomonas* sp. Is79 being the only exception. Another potential NO oxidase candidate is NirK, which is a reverse-operating nitrite reductase that is responsible for the reduction of nitrite (NO_2^-) into NO. NirK reportedly assists in the efficient oxidation of NH_3 into NO_2^- in AOB (Zorz et al., 2018).

Cytochromes are also involved in the energy metabolism of AOB. Cytochrome aa3, cytochrome bc1, cytochrome c_m-552, and cytochrome c-554 have been detected in different levels in various AOB (Zorz et al., 2018). Cytochrome c-554, the physiological electron acceptor in HAO, plays an important role in bacterial ammonia oxidation by mediating the electron transfer to the membrane-bound cytochrome c_m-552 located in the energy-generating respiratory system (Guo et al., 2013).

4.2 Molecular Detection and Identification of Ammonia-Oxidizing Bacteria

The specific molecular determination of AOB and their subsequent classification is based on the exploitation of conventional markers, such as 16S rDNA (Calvo & Garcia-Gil, 2004). Most of the knowledge regarding the ecology of ammonia oxidizers was obtained by examining the 16S rDNA gene and the *amoA* gene, the latter responsible for encoding the AmoA subunit of ammonia monooxygenase, which is a key enzyme that is shared by all three groups of ammonia oxidizers (Lehtovirta-Morley, 2018). The traditional method of analyzing AOB through isolation followed by bacterial culture is time-consuming and the AOB biomass yields are low. Moreover, it is difficult to remove the other heterotrophic contaminants from the medium when using this approach. However, the use of polymerase chain reaction (PCR) to amplify the 16S rDNA fragments followed by nucleotide sequencing has emerged as a better approach to investigate the phylogenetic affinities of the members of natural communities without the requirement of laboratory culture. Recently, denatured gradient gel electrophoresis (DGGE) of PCR amplicons of the 16S rDNA gene was performed to examine the AOB populations. This approach allowed for a further comprehensive and precise assessment of the sequence diversity and distributions in natural and anthropogenic environments. In addition, molecular methods, such as fluorescence *in-situ* hybridization (FISH), micro-auto-radiography–FISH (FISH–MAR), and a combination of isotope array and microelectrodes with FISH, have been applied widely for the detection of AOB in different environments, identification of novel strains of ammonia-oxidizing microorganisms, and analysis of the physiology and ecological interactions of uncultured AOB and other organisms (Guo et al., 2013; Kumwimba & Meng, 2019).

Culture-independent molecular methods, such as denaturing gradient gel electrophoresis (DGGE), real-time quantitative PCR (qPCR), and construction and analysis of 16S rDNA gene libraries, have also contributed to the identification of AOB and other communities in various habitats. The 16S rDNA gene is applied widely for the study of AOB phylogeny. The *amoA* gene is used frequently as a conventional functional marker for investigation regarding ammonia-oxidizing microorganisms. Real-time PCR evaluations and DGGE are used for assessing the diversity of AOB and other communities in various habitats. In addition, DGGE of the PCR amplicons of the 16S rDNA gene was performed to investigate the abundance and composition of AOB communities (Kumwimba & Meng, 2019).

Since the oxidation of ammonia is catalyzed by the enzyme AMO, which is encoded by the AMO gene, the *amoA* gene is frequently used as a molecular biomarker for determining the abundance and diversity of ammonia oxidizers. Certain recent studies also assessed the activity of AOB by determining the abundance of the *amoA* gene in the soil (Gao et al., 2018; Kumwimba & Meng, 2019; Simon et al., 2020). However, the use of *amoA* as a marker has certain limitations, such as its inability to target the γ-subclass of AOB. Nonetheless, this biomarker provides a better estimate of the actual numbers and activities of AOB owing to its higher sensitivity to genetic variation compared to that of the 16S rDNA-based probes. Therefore, scholars have proposed to use a combination of the 16S rDNA approach and *amoA* for targeting genes most reliably when analyzing the natural populations of AOB. Other studies have demonstrated the use of the *amoB* gene as a suitable molecular marker for AOB, reporting convenient detection of the AOB belonging to both β and γ subclasses of *Proteobacteria* using this marker gene (Guo et al., 2013).

The section ahead provides a summary of a few relevant molecular studies. Examples of the forward and reverse primers used in these studies are provided in Table 1.

The genes and the intergenic regions of the AmoCAB operon have been evaluated as potential molecular markers for analyzing the ammonia-oxidizing microorganisms

Table 1 Examples of the PCR primers used in a few molecular studies on AOB

Primer name	Sequence (5'-3')	Target	References
amoA	F: GGGGTTTCTACTGGTGGT	*amoA* gene	Zhang et al. (2011)
	R: CCCCTCKGSAAAGCCTTCTTC		
haoA	F: YTGYCAYAAYGGGRGYNGAYCAYAAYGAGT	HAO AB genes	Campbell et al. (2011)
haoB	R: ANNYGMYGRGMASYRTCCCACCA		
16S rDNA	27F: AGAGTTTGATCCTGGCTCAG	16S rDNA	Hayatsu et al. (2017)
	1492R: ACGGCTACCDTT-GTTACGACTT		
amoA	F: TTTGGTTGACTGGAAAGATCG	TAO100 *amoA*	
	R:GATACTGGGACAACATCCTTACC		
OCE	F: AGTGCGAATAGCATTGG	TAO100 16S rDNA	
	R: AACCCTCCCAACGTCTAGTTC		
16S rDNA	20F: GTAAT CGTCG GCCAG TAGAG TTTGA TCCTG GCTC	16S rDNA	Matsuno et al. (2013)
	1510R: CAGGA AACAG CTATG ACCGG CTACC TTGTT ACGAC		
CTO	189F: GGAGRAAAGCAGGGGATCG	16S rDNA	Sonthiphand and Limpiyakorn (2011)
	654R: CTAGCYTTGTAGTTTCAAACGC		
amoA	F: GGHGACTCCCAYTTCTGG	*amoA* genes	He et al. (2007)
	R: CCTCKGSAAAGCCTTCTTC		

belonging to Betaproteobacteria. The *amoB* gene was expected to perform better as a molecular marker, compared to the most commonly used *amoA* gene, to resolve closely-related species of AOB. However, various analyses conducted to evaluate the feasibility of using amoCAB, such as PCR, gradient gel electrophoresis, cloning, and sequencing, revealed that all amoCAB genes were equally efficient candidates for use as molecular markers to study AOB. Among all amoCAB genes, the amoC and amoA have been recommended for use as molecular markers for the study of less related species of AOB in environmental samples, while the amoB gene might serve better as a molecular marker to resolve the closely related species of AOB due to lower sequence conservation (Junier et al., 2009).

Yamamoto et al. (2010) amplified the bacterial *amoA* gene through PCR using the amoA1F/amoA2R primer set. The objective of the study was to determine the diversity and abundance of AOB using the denaturing gradient gel electrophoresis and real-time PCR techniques. In 2011, Zhang et al. determined the numbers of the AOB amoA gene by performing PCR using the primers amoA-1F and amoA-2R. In this study, the diversity of AOB in six bioreactors of different wastewater treatment plants (WWTPs) located in four countries was investigated using the *amoA* and 16S rDNA genes as biomarkers. High-throughput pyrosequencing of the 16S rDNA genes was performed to tabulate the diversity of AOB and their relative abundance in the whole bacterial community. The results of the study revealed that the AOB amoA gene varied diversely among different wastewater treatment plants, while the 16S rRNA gene was relatively conserved. In 2011, Campbell et al. used the degenerate primer pair haoA-F and haoB-Rev1, which was designed for the gammaproteobacterial detection using PCR, for the amplification of the haoAB genes and genomic DNA. In addition, the sequence similarities between the AOB-specific proteins AmoA and HaoA and those among the rrsA, gyrB, and rpoD genes were investigated using the NCBI blast program. Another study reported in 2011 was the one conducted by Sonthiphand and Limpiyakorn, in which the bacterial *amoA* genes were quantified using primers amoA 1F and amoA 2R, while the AOB 16S rRNA gene was quantified using primers CTO 189A/Bf, CTO189Cf, and CTO 654r. According to the results, all the analyzed sequences obtained from the samples were classified into four clusters—unknown *Nitrosomonas* cluster, *Nitrosomonas europaea* cluster, *Nitrosomonas communis* cluster, and *Nitrosomonas oligotropha* cluster.

Fujitani et al. (2015) isolated novel pure strains of AOB belonging to the *Nitrosomonas mobilis* species from autotrophic nitrification granules. Cell growth was evaluated through FISH and fluorescence microscopy, while nitrite production was confirmed using the Griess reagent. Pure bacterial strains were identified through the 16S rRNA and *amoA* gene sequence analysis. The 16S rRNA gene fragment of the total DNA was amplified using the 27f/1492r primer set. The *amoA* gene fragment of the total DNA was amplified using the amoA1F/amoA2R bacterial primer set. Urakawa et al. (2015) isolated a pure culture of spiral-shaped, Gram-negative, chemolithotrophic, AOB, which the authors designated as APG3, from the sandy lake sediment collected from Green Lake, Seattle, USA. The phylogenetic analyses based on the 16S rRNA gene sequence conducted for this APG3 strain revealed *Nitrosospira multiformis* (cluster 3) as its closest species and that it belonged to cluster 0 of the *Nitrosospira* genus, a cluster not represented currently by any of the already described species. Since the general polyphasic taxonomic analysis revealed that the APG3T strain represents a novel species belonging to genus *Nitrosospira*, it is proposed that this strain be recognized as a novel species. Hayatsu et al. (2017) evaluated the enriched DNA fragments of AOB primers 27F and 1492R—from the bacteria-specific 16S rRNA gene sequences using a FISH probe, PCR, and quantitative PCR (qPCR). The study aimed at the isolation and characterization of an acid-adapted AOB from acidic agricultural soil. In order to genomically validate the taxonomic and evolutionary relationships between the known related AOBs, the total bacterial 16S rRNA gene copy number was quantified by performing qPCR using a bacteria-specific primer set. The TAO100 amoA gene was amplified using the primers TAO-amoAF and TAO-amoAR. It was concluded that the TAO100 strain of γ-AOB was uniquely adapted to acidic conditions and could contribute significantly to the nitrification process in acidic soils. In 2020, Lukumbuzya et al. developed and experimentally evaluated a novel set of highly specific 16S rRNA-targeted oligonucleotide probes that could complement the existing probes and thereby enable the specific detection and differentiation of the known major phylogenetic clusters of the AOB belonging to Betaproteobacteria. The developed novel probes have been successfully applied to visualize AOB and measure their concentrations in activated sludge and biofilm specimens from seven pilot and full-scale wastewater treatment systems. The applicability of the novel probe set was evaluated in FISH experiments conducted with nitrification-activated sludge samples from different WWTPs in Austria, Germany, and Denmark. The results revealed that the novel AOB belonging to Betaproteobacteria probe mix is suitable for detecting and measuring β-AOB in AATs. Importantly, the probe mix offered better specificity for β-AOB compared to the previously reported probes.

5 Conclusion

AOB play a fundamental role in the nitrogen cycle in terrestrial, marine, and industrial systems. The significant role of nitrification in both natural environments and industrial systems such as wastewater treatment systems renders it important to decipher and understand the physiology of the microorganisms involved in the nitrification process. In this context, the present report discussed the general characteristics, the types, and the key enzymes of AOB, which are responsible for the oxidation of ammonia to nitrite, the first step in the nitrification process.

References

Arp, D. J., Chain, P. S., & Klotz, M. G. (2007). The impact of genome analyses on our understanding of ammonia-oxidizing bacteria. *Annual Review of Microbiology, 61*, 503–528. https://doi.org/10.1146/annurev.micro.61.080706.093449

Bai, Y., Sun, Q., Wen, D., & Tang, X. (2012). Abundance of ammonia-oxidizing bacteria and archaea in industrial and domestic wastewater treatment systems. *FEMS Microbiology Ecology, 80*(2), 323–330. https://doi.org/10.1111/j.1574-6941.2012.01296.x

Bernhard, A. E., & Bollmann, A. (2010). Estuarine nitrifiers: New players, patterns and processes. *Estuarine, Coastal and Shelf Science, 88*(1), 1–11. https://doi.org/10.1016/j.ecss.2010.01.023

Bollmann, A., Sedlacek, C. J., Norton, J., Laanbroek, H. J., Suwa, Y., Stein, L. Y., Klotz, M. G., Arp, D., Sayavedra-Soto, L., Lu, M., Bruce, D., Detter, C., Tapia, R., Han, J., Woyke, T., Lucas, S. M., Pitluck, S., Pennacchio, L., Nolan, M., Land, M. L., Huntemann, M., Deshpande, S., Han, C., Chen, A., Kyrpides, N., Mavromatis, K., Markowitz, V., Szeto, E., Ivanova, N., Mikhailova, N., Pagani, I., Pati, A., Peters, L., Ovchinnikova, G., & Goodwin, L. A. (2013). Complete genome sequence of Nitrosomonas sp. Is79, an ammonia oxidizing bacterium adapted to low ammonium concentrations. *Standards in Genomic Sciences, 7*(3), 469–482. https://doi.org/10.4056/sigs.3517166

Calvó, L., & Garcia-Gil, L. J. (2004). Use of amoB as a new molecular marker for ammonia-oxidizing bacteria. *Journal of Microbiological Methods, 57*(1), 69–78. https://doi.org/10.1016/j.mimet.2003.11.019

Campbell, M. A., Chain, P. S., Dang, H., El Sheikh, A. F., Norton, J. M., Ward, N. L., Ward, B. B., & Klotz, M. G. (2011). Nitrosococcus watsonii sp. nov., a new species of marine obligate ammonia-oxidizing bacteria that is not omnipresent in the world's oceans: calls to validate the names '*Nitrosococcus halophilus*' and '*Nitrosomonas mobilis*'. *FEMS Microbiology Ecology, 76*(1), 39–48. https://doi.org/10.1111/j.1574-6941.2010.01027.x

Caranto, J. D., & Lancaster, K. M. (2017). Nitric oxide is an obligate bacterial nitrification intermediate produced by hydroxylamine oxidoreductase. *Proceedings of the National Academy of Sciences, 114*(31), 8217–8222. https://doi.org/10.1073/pnas.1704504114

Chu, H., Fujii, T., Morimoto, S., Lin, X., Yagi, K., Hu, J., & Zhang, J. (2007). Community structure of ammonia-oxidizing bacteria under long-term application of mineral fertilizer and organic manure in a sandy loam soil. *Applied and Environmental Microbiology, 73*(2), 485–491. https://doi.org/10.1128/AEM.01536-06

Daims, H., Lebedeva, E. V., Pjevac, P., Han, P., Herbold, C., Albertsen, M., Jehmlich, N., Palatinszky, M., Vierheilig, J., Bulaev, A., Kirkegaard, R. H., von Bergen, M., Rattei, T., Bendinger, B., Nielsen, P. H., & Wagner, M. (2015). Complete nitrification by Nitrospira bacteria. *Nature, 528*(7583), 504–509. https://doi.org/10.1038/nature16461

Daims, H., Lücker, S., & Wagner, M. (2016). A new perspective on microbes formerly known as nitrite-oxidizing bacteria. *Trends in Microbiology, 24*(9), 699–712. https://doi.org/10.1016/j.tim.2016.05.004

Díaz, A. I., Oulego, P., Laca, A., González, J. M., & Díaz, M. (2019). Metagenomic analysis of bacterial communities from a nitrification–denitrification treatment of landfill leachates. *CLEAN–Soil, Air, Water, 47*(11), 1900156. https://doi.org/10.1002/clen.201900156

Ding, S., Zheng, P., Lu, H., Chen, J., Mahmood, Q., & Abbas, G. (2013). Ecological characteristics of anaerobic ammonia oxidizing bacteria. *Applied Microbiology and Biotechnology, 97*(5), 1841–1849. https://doi.org/10.1007/s00253-013-4697-0

Dong, Y., Zhang, Y., Tu, B., & Miao, J. (2014). Immobilization of ammonia-oxidizing bacteria by calcium alginate. *Ecological Engineering, 73*, 809–814. https://doi.org/10.1016/j.ecoleng.2014.09.020

Esen, E. (2013). Investigation of microbial communities in laboratory scale anammox reactors using molecular methods. İstanbul Technical University, Institute of Science and Technology, Thesis (M.Sc.), p. 45.

Fujitani, H., Kumagai, A., Ushiki, N., Momiuchi, K., & Tsuneda, S. (2015). Selective isolation of ammonia-oxidizing bacteria from autotrophic nitrifying granules by applying cell-sorting and sub-culturing of microcolonies. *Frontiers in Microbiology, 6*, 1159. https://doi.org/10.3389/fmicb.2015.01159

Gao, S., Cao, W., Zou, C., Gao, J., Huang, J., Bai, J., Zeng, N., Shimizu, K., Wright, A., & Dou, F. (2018). Ammonia-oxidizing archaea are more sensitive than ammonia-oxidizing bacteria to long-term application of green manure in red paddy soil. *Applied Soil Ecology, 124*, 185–193. https://doi.org/10.1016/j.apsoil.2017.09.041

Garcia, J. C., Urakawa, H., Le, V. Q., Stein, L. Y., Klotz, M. G., & Nielsen, J. L. (2013). Draft genome sequence of Nitrosospira sp. strain APG3, a psychrotolerant ammonia-oxidizing bacterium isolated from sandy lake sediment. *Genome Announcements, 1*(6), e00930–13. https://doi.org/10.1128/genomeA.00930-13

Guo, J., Peng, Y., Wang, S., Ma, B., Ge, S., Wang, Z., Huang, H., Zhang, J., & Zhang, L. (2013). Pathways and organisms involved in ammonia oxidation and nitrous oxide emission. *Critical Reviews in Environmental Science and Technology, 43*(21), 2213–2296. https://doi.org/10.1080/10643389.2012.672072

Hayatsu, M., Tago, K., Uchiyama, I., Toyoda, A., Wang, Y., Shimomura, Y., Okubo, T., Kurisu, F., Hirono, Y., Nonaka, K., Akiyama, H., Itoh, T., & Takami, H. (2017). An acid-tolerant ammonia-oxidizing γ-proteobacterium from soil. *The ISME Journal, 11*(5), 1130–1141. https://doi.org/10.1038/ismej.2016.191

He, J. Z., Shen, J. P., Zhang, L. M., Zhu, Y. G., Zheng, Y. M., Xu, M. G., & Di, H. (2007). Quantitative analyses of the abundance and composition of ammonia-oxidizing bacteria and ammonia-oxidizing archaea of a Chinese upland red soil under long-term fertilization practices. *Environmental Microbiology, 9*(9), 2364–2374. https://doi.org/10.1111/j.1462-2920.2007.01358.x

Holmes, D. E., Dang, Y., & Smith, J. A. (2019). Nitrogen cycling during wastewater treatment. *Advances in Applied Microbiology, 106*, 113–192. https://doi.org/10.1016/bs.aambs.2018.10.003

Junier, P., Kim, O. S., Junier, T., Ahn, T. S., Imhoff, J. F., & Witzel, K. P. (2009). Community analysis of betaproteobacterial ammonia-oxidizing bacteria using the amoCAB operon. *Applied Microbiology and Biotechnology, 83*(1), 175–188. https://doi.org/10.1007/s00253-009-1923-x

Keluskar, R., Nerurkar, A., & Desai, A. (2013). Mutualism between autotrophic ammonia-oxidizing bacteria (AOB) and heterotrophs present in an ammonia-oxidizing colony. *Archives of Microbiology, 195*(10), 737–747. https://doi.org/10.1007/s00203-013-0926-2

Kim, S. W., Miyahara, M., Fushinobu, S., Wakagi, T., & Shoun, H. (2010). Nitrous oxide emission from nitrifying activated sludge dependent on denitrification by ammonia-oxidizing bacteria. *Bioresource Technology, 101*(11), 3958–3963. https://doi.org/10.1016/j.biortech.2010.01.030

Klotz, M. G., Arp, D. J., Chain, P. S., El-Sheikh, A. F., Hauser, L. J., Hommes, N. G., Larimer, F. W., Malfatti, S. A., Norton, J. M., Poret-Peterson, A. T., Vergez, L. M., & Ward, B. B. (2006). Complete genome sequence of the marine, chemolithoautotrophic, ammonia-oxidizing bacterium Nitrosococcus oceani ATCC 19707. *Applied and Environmental Microbiology, 72*(9), 6299–6315. https://doi.org/10.1128/AEM.00463-06

Kozlowski, J. A., Kits, K. D., & Stein, L. Y. (2016a). Comparison of nitrogen oxide metabolism among diverse ammonia-oxidizing bacteria. *Frontiers in Microbiology, 7*, 1090. https://doi.org/10.3389/fmicb.2016.01090

Kozlowski, J. A., Kits, K. D., & Stein, L. Y. (2016b). Complete genome sequence of Nitrosomonas ureae strain Nm10, an Oligotrophic Group 6a Nitrosomonad. *Genome Announcements, 4*(2), e00094–16. https://doi.org/10.1128/genomeA.00094-16

Kozlowski, J. A., Kits, K. D., & Stein, L. Y. (2016c). Genome sequence of Nitrosomonas communis strain Nm2, a mesophilic ammonia-oxidizing bacterium isolated from Mediterranean soil. *Genome Announcements, 4*(1), e01541–15. https://doi.org/10.1128/genomeA.01541-15

Kozlowski, J. A., Price, J., & Stein, L. Y. (2014). Revision of N_2O-producing pathways in the ammonia-oxidizing bacterium Nitrosomonas europaea ATCC 19718. *Applied and Environmental Microbiology, 80*(16), 4930–4935. https://doi.org/10.1128/AEM.01061-14

Kumwimba, M. N., & Meng, F. (2019). Roles of ammonia-oxidizing bacteria in improving metabolism and cometabolism of trace organic chemicals in biological wastewater treatment processes: A review. *Science of the Total Environment, 659*, 419–441. https://doi.org/10.1016/j.scitotenv.2018.12.236

Lehtovirta-Morley, L. E. (2018). Ammonia oxidation: Ecology, physiology, biochemistry and why they must all come together. *FEMS Microbiology Letters, 365*(9), fny058. https://doi.org/10.1093/femsle/fny058

Lukumbuzya, M., Kristensen, J. M., Kitzinger, K., Pommerening-Röser, A., Nielsen, P. H., Wagner, M., Daims, H., & Pjevac, P. (2020). A refined set of rRNA-targeted oligonucleotide probes for in situ detection and quantification of ammonia-oxidizing bacteria. *Water Research, 186*, 116372. https://doi.org/10.1016/j.watres.2020.116372

Matsuno, T., Horii, S., Sato, T., Matsumiya, Y., & Kubo, M. (2013). Analysis of nitrification in agricultural soil and improvement of nitrogen circulation with autotrophic ammonia-oxidizing bacteria. *Applied Biochemistry and Biotechnology, 169*(3), 795–809. https://doi.org/10.1007/s12010-012-0029-6

Mukhtar, H., Lin, Y. P., & Anthony, J. (2017). Ammonia oxidizing archaea and bacteria in east Asian paddy soils—A mini review. *Environments, 4*(4), 84. https://doi.org/10.3390/environments4040084

Norton, J. M. (2011). Diversity and environmental distribution of ammonia-oxidizing bacteria. *Nitrification*, 39–55. https://doi.org/10.1128/9781555817145.ch3

Norton, J. M., Klotz, M. G., Stein, L. Y., Arp, D. J., Bottomley, P. J., Chain, P. S., Hauser, L. J., Land, M. L., Larimer, F. W., Shin, M. W., & Starkenburg, S. R. (2008). Complete genome sequence of Nitrosospira multiformis, an ammonia-oxidizing bacterium from the soil environment. *Applied and Environmental Microbiology, 74*(11), 3559–3572. https://doi.org/10.1128/AEM.02722-07

Osburn, E. D., & Barrett, J. E. (2020). Abundance and functional importance of complete ammonia-oxidizing bacteria (comammox) versus canonical nitrifiers in temperate forest soils. *Soil Biology and Biochemistry, 145*, 107801. https://doi.org/10.1016/j.soilbio.2020.107801

Park, H. D., & Noguera, D. R. (2007). Characterization of two ammonia-oxidizing bacteria isolated from reactors operated with low dissolved oxygen concentrations. *Journal of Applied Microbiology, 102*(5), 1401–1417. https://doi.org/10.1111/j.1365-2672.2006.03176.x

Purkhold, U., Pommerening-Röser, A., Juretschko, S., Schmid, M. C., Koops, H. P., & Wagner, M. (2000). Phylogeny of all recognized species of ammonia oxidizers based on comparative 16S rRNA and amoA sequence analysis: implications for molecular diversity surveys. *Applied and Environmental Microbiology, 66*(12), 5368–5382. https://doi.org/10.1128/AEM.66.12.5368-5382.2000

Rice, M. C., Norton, J. M., Valois, F., Bollmann, A., Bottomley, P. J., Klotz, M. G., Laanbroek, H. J., Suwa, Y., Stein, L. Y., Sayavedra-Soto, L., Woyke, T., Shapiro, N., Goodwin, L. A., Huntemann, M., Clum, A., Pillay, M., Kyrpides, N., Varghese, N., Mikhailova, N., … Daum, C. (2016). Complete genome of Nitrosospira briensis C-128, an ammonia-oxidizing bacterium from agricultural soil. *Standards in Genomic Sciences, 11*(1), 1–8. https://doi.org/10.1186/s40793-016-0168-4

Sarı, T., & Mertoğlu, B. (2019). Anaerobik amonyum oksidasyonu (Anammox). *Uludağ University, Journal of The Faculty of Engineering, 24*(3), 523–548. https://doi.org/10.17482/uumfd.502482

Schebor, H. A. (2014). *Ammonia-oxidizing bacteria and archaea across a freshwater trophic gradient* (p. 67). Miami University.

Schmidt, I., Steenbakkers, P. J., op den Camp, H. J., Schmidt, K., & Jetten, M. S. (2004). Physiologic and proteomic evidence for a role of nitric oxide in biofilm formation by Nitrosomonas europaea and other ammonia oxidizers. *Journal of Bacteriology, 186*(9), 2781–2788. https://doi.org/10.1128/JB.186.9.2781-2788.2004

Sedlacek, C. J., McGowan, B., Suwa, Y., Sayavedra-Soto, L., Laanbroek, H. J., Stein, L. Y., Norton, J. M., Klotz, M. G., & Bollmann, A. (2019). A physiological and genomic comparison of Nitrosomonas cluster 6a and 7 ammonia-oxidizing bacteria. *Microbial Ecology, 78*(4), 985–994. https://doi.org/10.1007/s00248-019-01378-8

Sedlacek, C. J., Nielsen, S., Greis, K. D., Haffey, W. D., Revsbech, N. P., Ticak, T., Laanbroek, H. J., & Bollmann, A. (2016). Effects of bacterial community members on the proteome of the ammonia-oxidizing bacterium Nitrosomonas sp. strain Is79. *Applied and Environmental Microbiology, 82*(15), 4776–4788. https://doi.org/10.1128/AEM.01171-16

Shah, M. P. (2020). *Microbial bioremediation and biodegradation*. Springer.

Shah, M. P. (2021). *Removal of refractory pollutants from wastewater treatment plants*. CRC Press.

Simon, P., De Klein, C., Gerard, E., & Shi, S. (2020). Ammonia-oxidizing bacteria dynamics affected by plantain under synthetic cattle urine patches. *Journal of New Zealand Grasslands, 82*, 97–102. https://doi.org/10.33584/jnzg.2020.82.433

Soliman, M., & Eldyasti, A. (2018). Ammonia-Oxidizing Bacteria (AOB): Opportunities and applications—a review. *Reviews in Environmental Science and Bio/technology, 17*(2), 285–321. https://doi.org/10.1007/s11157-018-9463-4

Sonthiphand, P., & Limpiyakorn, T. (2011). Change in ammonia-oxidizing microorganisms in enriched nitrifying activated sludge. *Applied Microbiology and Biotechnology, 89*(3), 843–853. https://doi.org/10.1007/s00253-010-2902-y

Stein, L. Y., Arp, D. J., Berube, P. M., Chain, P. S., Hauser, L., Jetten, M. S., Klotz, M. G., Larimer, F. W., Norton, J. M., den Camp, H. J. M., Shin, M., & Wei, X. (2007). Whole-genome analysis of the ammonia-oxidizing bacterium, Nitrosomonas eutropha C91: implications for niche adaptation. *Environmental Microbiology, 9*(12), 2993–3007. https://doi.org/10.1111/j.1462-2920.2007.01409.x

Stein, L. Y. (2019). Insights into the physiology of ammonia-oxidizing microorganisms. *Current Opinion in Chemical Biology, 49*, 9–15. https://doi.org/10.1016/j.cbpa.2018.09.003

Stein, L. Y., Campbell, M. A., & Klotz, M. G. (2013). Energy-mediated vs. ammonium-regulated gene expression in the obligate ammonia-oxidizing bacterium, Nitrosococcus oceani. *Frontiers in Microbiology, 4*, 277. https://doi.org/10.3389/fmicb.2013.00277

Thandar, S. M., Ushiki, N., Fujitani, H., Sekiguchi, Y., & Tsuneda, S. (2016). Ecophysiology and comparative genomics of Nitrosomonas mobilis Ms1 isolated from autotrophic nitrifying granules of wastewater treatment bioreactor. *Frontiers in Microbiology, 7*, 1869. https://doi.org/10.3389/fmicb.2016.01869

Urakawa, H., Garcia, J. C., Nielsen, J. L., Le, V. Q., Kozlowski, J. A., Stein, L. Y., Lim, C. K., Pommerening-Röser, A., Martens-Habbena, W., Stahl, D. A., & Klotz, M. G. (2015). Nitrosospira lacus sp. nov., a psychrotolerant, ammonia-oxidizing bacterium from sandy lake sediment. *International Journal of Systematic and Evolutionary Microbiology, 65*(Pt_1), 242–250. https://doi.org/10.1099/ijs.0.070789-0

Wang, C., Tang, S., He, X., & Ji, G. (2020). The abundance and community structure of active ammonia-oxidizing archaea and ammonia-oxidizing bacteria shape their activities and contributions in coastal wetlands. *Water Research, 171*, 115464. https://doi.org/10.1016/j.watres.2019.115464

Wang, Y., Ma, L., Mao, Y., Jiang, X., Xia, Y., Yu, K., Li, B., & Zhang, T. (2017). Comammox in drinking water systems. *Water Research, 116*, 332–341. https://doi.org/10.1016/j.watres.2017.03.042

Ward, L. M., Johnston, D. T., & Shih, P. M. (2021). Phanerozoic radiation of ammonia oxidizing bacteria. *Scientific Reports, 11*(1), 1–9. https://doi.org/10.1038/s41598-021-81718-2

Yamamoto, N., Otawa, K., & Nakai, Y. (2010). Diversity and abundance of ammonia-oxidizing bacteria and ammonia-oxidizing archaea during cattle manure composting. *Microbial Ecology, 60*(4), 807–815. https://doi.org/10.1007/s00248-010-9714-6

Zhang, T., Ye, L., Tong, A. H. Y., Shao, M. F., & Lok, S. (2011). Ammonia-oxidizing archaea and ammonia-oxidizing bacteria in six full-scale wastewater treatment bioreactors. *Applied Microbiology and Biotechnology, 91*(4), 1215–1225. https://doi.org/10.1007/s00253-011-3408-y

Zorz, J. K., Kozlowski, J. A., Stein, L. Y., Strous, M., & Kleiner, M. (2018). Comparative proteomics of three species of ammonia-oxidizing bacteria. *Frontiers in Microbiology, 9*, 938. https://doi.org/10.3389/fmicb.2018.00938

Anaerobic Ammonium Oxidation Using Engineered Nanomaterials as Potential Modulators

Sougata Ghosh, Bishwarup Sarkar, and Sirikanjana Thongmee

Abstract Anaerobic ammonium oxidation (anammox) process-mediated nitrogen removal technology is an effective and sustainable method for wastewater treatment. Extensive and large-scale utilization of anammox bacteria is still not feasible because of its long doubling time as well as vulnerability to diverse operating conditions. In order to alter the activity of anammox bacteria, engineered nanomaterials are used as external modulators. Therefore, this chapter elaborates on the effects of different engineered nanomaterials on anammox-based biological nitrogen removal systems. Nanostructures, such as silver nanoparticles, copper nanoparticles, nano zero-valent iron, are reported to stimulate the overall growth of *Candidatus Jettenia, Candidatus Kuenenia, Candidatus Brocadia, Candidatus Scalindua,* and others thereby enhancing total nitrogen removal rate. Likewise, nanostructured copper oxide, iron oxide, magnesium oxide, manganese oxide, titanium dioxide, zinc oxide can stimulate the biomass generation in anammox bacteria like *Candidatus Anammoxoglobus, Parcubacteria_genera_incertae_sedis,* and *Denitratisoma.* Various carbon-based nanostructures such as graphene nanosheets, γ-Fe_2O_3 nanoparticles, and reduced graphene oxide can also modulate the anammox process effectively. Nanomaterials can augment anammox-related enzymes such as nitrite oxidoreductase, hydrazine dehydrogenase, and hydrazine synthase. Expression of the vital genes such as *hzsA*, *nirS*, and *hdh* can be modulated by the nanoparticles which in turn influence the overall efficiency of the anammox process. Additionally, treatment with nanoparticles can enhance exopolysaccharide production, quorum-sensing, and cyclic diguanylate (c-di-GMP) level that can offer better cell signaling and attachment of the bacterial biomass. Further exploration of the reaction kinetics would provide better understanding of the mechanism of interaction between these engineered nanomaterials and anammox bacteria that can revolutionize wastewater treatment.

S. Ghosh (✉) · S. Thongmee
Department of Physics, Faculty of Science, Kasetsart University, Bangkok, Thailand
e-mail: ghoshsibb@gmail.com

S. Ghosh
Department of Microbiology, School of Science, RK. University, Rajkot, Gujarat, India

B. Sarkar
College of Science, Northeastern University, Boston, MA, USA

M. P. Shah (ed.), *Anammox Technology in Industrial Wastewater Treatment*,
https://doi.org/10.1007/978-981-99-3459-1_3

Keywords Anaerobic ammonium oxidation · Metal nanoparticles · Metal oxide · Graphene nanosheets · Enzymes · Total nitrogen removal rate

1 Introduction

The discovery of autotrophic anaerobic ammonium oxidation (anammox) bacteria has provided an effective way for the wastewater treatment process, which is economical and efficient (Li et al., 2018a). Hence, various anammox processes are developed for the treatment of industrial wastewater containing different amounts of ammonium and nitrite. However, anammox bacteria has a considerable doubling time and is vulnerable to the varying operating conditions, which make it challenging to be utilized for the long-term and large-scale applications for wastewater treatment (Li et al., 2019).

With the continuous and rapid development of engineered nanomaterials, multiple industries are utilizing such nanomaterials for catalytic processes in manufacturing, electronics, and effluent treatment (Shende et al., 2017). Due to smaller dimension, nanoparticles have large surface area and attractive physicochemical and optoelectronic properties. The catalytic and biological properties of nanoparticles are dependent on their shape and size (Bloch et al., 2021). During industrial processes involving nanotechnology, these nanomaterials are occasionally released into the environment which further interacts with the aquatic microflora (Li et al., 2019). The application of such nanomaterials and nanoparticles (NPs) has been reported to exhibit both positive and negative impacts on the anammox activity under different conditions. Hence, this chapter elaborates on various studies which investigated the activity of nanomaterials on the anammox property. Several metallic and metal-oxide NPs such as silver, copper, zinc oxide, maghemite, zero-valent iron, magnesium oxide, titanium oxide, and manganese oxide have been demonstrated to alter the anammox activity that is summarized in Table 1. In addition, non-metallic nanomaterials such as graphene nanosheets and reduced graphene oxide can enhance the anammox activity as well. Therefore, further studies on identifying the mechanism of action of these NPs would provide useful information on the application of certain NPs for enhancing the anammox activity during wastewater treatments.

2 Metallic Nanoparticles

A variety of engineered nanomaterials are reported to alter the anammox activity which in turn, modulates the biological nitrogen removal processes. In one such study, silver nanoparticles (AgNPs) promoted the proliferation of anammox bacteria which in turn, improved the nitrogen removal ability (Peng et al., 2019). The nitrogen removal rate (NRR) was significantly improved in the presence of 1.0 mM of AgNPs along with a 6.5-fold increase in the abundance of *Brocadiaceae* family. Redundancy

Table 1 Effect of various nanoparticles in the anammox process

Type of nanomaterial	Nanomaterial concentration	Effect on the nitrogen removal process	Reference
Metallic NPs			
AgNPs	1.0 mM	Enhancement of NRR along with formation of encapsulins by the anammox bacteria	Peng et al. (2019)
CuNPs	5.0 mg/L	Decrease in NRE by 55% while the anammox biomass showed gradual resilience with 21.4% increase in the EPS content	Cheng et al. (2020)
nZVI	50 mg/L	Improvement in NRE by 82.44% in by promotion of the quorum-sensing of anammox bacteria	Wang et al. (2021)
nZVI	5000 ppb	Improvement of both ammonium and nitrite removal rates due to the formation of cavity in the anammox granules	Erdim et al. (2019)
Metal oxide NPs			
CuONPs	5 mg of Cu/L	Inhibitory action due to long-term self-adaptability of anammox bacteria	Zhang et al., (2018a)
Fe_3O_4NPs	10 mg/L	Improvement of NRE by 77.5% due to release of Fe ions by Fe_3O_4NPs that improved the anammox activity	Li et al., (2018b)
Fe_3O_4NPs	200 mg/L	Enhancement of SAA and metabolism of the anammox bacteria	Xu et al., (2020)
MgONPs	200 mg/L	Reduction in NRE that could be alleviated by 5 mg/L of BSA due to NPs adsorption	Ma et al., (2022)
MHNPs	200 mg/L	Enhancement of SAA	Zhang et al., (2018b)
MnO_2NPs	200 mg/L	Improvement of NRE by 93.1% along with increased EPS content, NLR as well as SAA	Xu et al., (2019)
TiO_2NPs	300 mg/L	81% inhibition of the nitrogen removal ability through adsorption of the surface of anammox bacteria	Tuğba et al., (2020)
ZnONPs	10 mg/L	Reduction in the nitrogen removal capacity by 90% after the first shock which was gradually alleviated after five cycles	Zhang et al., (2018c)
ZnONPs	-	Severe reduction in anammox nitrogen removal ability with an IC_{50} value of 11.7 ± 0.9 mg g/VSS	Song et al., (2018)

(continued)

Table 1 (continued)

Type of nanomaterial	Nanomaterial concentration	Effect on the nitrogen removal process	Reference
ZnONPs	150 mg/L	Reduction of anammox activity by 64.0%	Zhao et al., (2019)
Carbon-based nanomaterials			
GNs	10 mg/L	Enhancement of NRR by 55%	Elreedy et al., (2021)
GO	0.1 g/L	Improvement of anammox activity by 10.26%	Wang et al., (2013)
rGO	-	Enhancement of anammox activity by 17%	Tomaszewski et al., (2019)

analysis then revealed the dominance of *Candidatus Jettenia* when 1.0 mM of AgNPs was introduced into the system. Transmission electron microscopy (TEM) images further highlighted the presence of special "cavity" like structures that were termed encapsulins which were suggested to confer NP resistance to the anammox bacteria as evident from Fig. 1. X-ray tomography images further implied an increase in the porosity from 63.08 $\pm$ 2.59% to 73.59 $\pm$ 3.02% as well as a higher abundance of pores with a size of 37.93 μm after the addition of 1.0 mM of AgNPs. Moreover, the expression of anammox-related enzymes such as nitrite oxidoreductase (NirS), hydrazine dehydrogenase (Hdh), and hydrazine synthase (HZS) was also upregulated in the presence of AgNPs.

Copper nanoparticles (CuNPs) are biocompatible with various therapeutic properties such as antidiabetic, antioxidant, and antimicrobial activities (Bhagwat et al., 2018). Cheng et al. (2020) demonstrated the resistance of anammox granules toward CuNPs that was enhanced with multiple shocks. In addition, oxytetracycline (OTC) was also supplemented in the process after 1 month of the reactor operation under normal conditions. The concentration of ammonia and nitrite in the effluents increased to 71.0 and 41.6 mg/L after the addition of 0.5 mg/L of OTC and CuNPs, respectively. Moreover, a 10% decrease in the total nitrogen removal efficiency (NRE) was also observed. Similarly, a 20% decrease in the total NRE was seen after the addition of 1 mg/L of both OTC and CuNPs. The nitrogen removal capacity was gradually reduced with increasing concentrations of CuNPs with a maximum decrease of around 55% of total NRE in the presence of 5.0 mg/L of CuNPs and OTC as compared to the control reactor without any supplementation of these toxic stimulators. However, the resistance of anammox bacteria was improved with subsequent exposure to CuNPs and OTC with more abundance of genes such as *hzsA*, *nirS*, and *hdh*. High-throughput 16S rRNA gene sequencing results also revealed an improved abundance of *Candidatus Kuenenia* with gradual acclimation to the CuNPs shock as evidenced by a 21.4% increase in the exopolysaccharide (EPS) content of the anammox biomass. Notably, the accumulation of Cu(II) ions released from the CuNPs in the anammox biomass was proposed to be the main mechanism of toxicity.

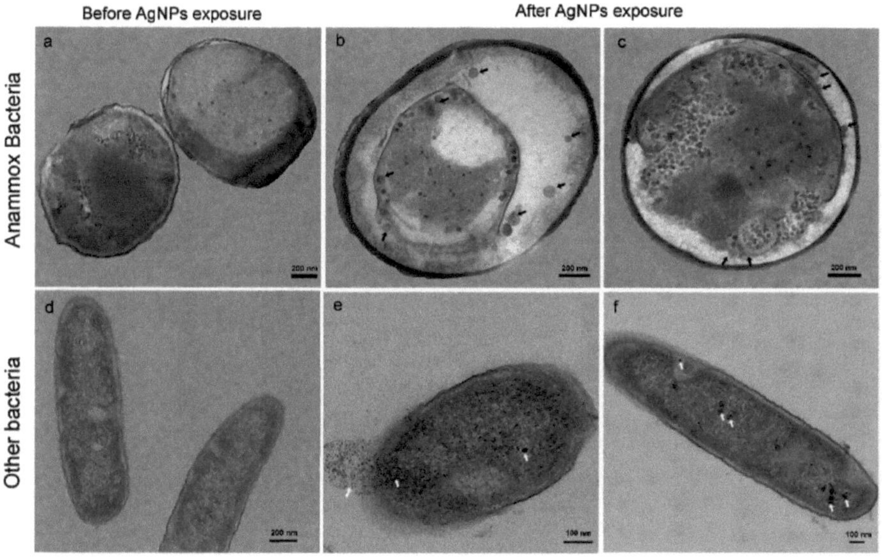

Fig. 1 TEM images of bacteria in anammox granules. TEM image of anammox bacteria before (**a**) and after exposure to AgNPs (**b**, **c**). TEM image of the remaining bacteria before (**d**) and after exposure to AgNPs (**e**, **f**). Reprinted with permission from Peng, M. W., Yu, X. L., Guan, Y., Liu, P., Yan, P., Fang, F., Guo, J., Chen, Y. P., 2019. Underlying promotion mechanism of high concentration of silver nanoparticles on anammox process. ACS Nano 13, 14,500–14,510. Copyright © 2019 American Chemical Society

Similarly, Wang et al. (2021) also recently reported the improvement of anammox activity in the presence of nano zero-valent iron (nZVI) by enhancing the quorum-sensing system of the anammox bacteria. The nitrogen removal capacity of nZVI was monitored wherein anammox activity was gradually increased from 15 to 65 days of incubation. A maximum ammonium removal efficiency of more than 85% was achieved with less than 10 mg of N/L was observed in the effluent after 87 days of incubation. Likewise, the nitrite removal efficiency (NRE) of 82.44% was obtained which was promoted by the addition of 50 mg/L of nZVI. In addition, higher concentrations of nZVI exhibited toxicity towards microbes hence, a low concentration of nZVI was supplemented in the UASB reactor. Moreover, 16S rRNA gene sequencing revealed the dominance of *Candidatus Brocadia* in the seed of anammox granular sludge wherein nZVI supplementation further increased the abundance of anammox bacteria in the reactor. The possible mechanism for promoting the anammox bacteria by the nanomaterial is illustrated in Fig. 2. Other anammox bacteria such as *Candidatus Brocadia fulgida*, *Candidatus Scalindua brodae*, *Candidatus Brocadia anammoxidans*, *Candidatus Jettenia asiatica*, and *Candidatus Anammoxoglobus propionicus*. Additionally, metagenomic sequencing of the microbial community indicated the increase in the abundance of signaling molecules such as cyclic diguanylate (c-di-GMP) from 148 targeted reads per million map reads (rpmr) to 252 rpmr. The increased concentration of c-di-GMP was suggested to immobilize the microbial

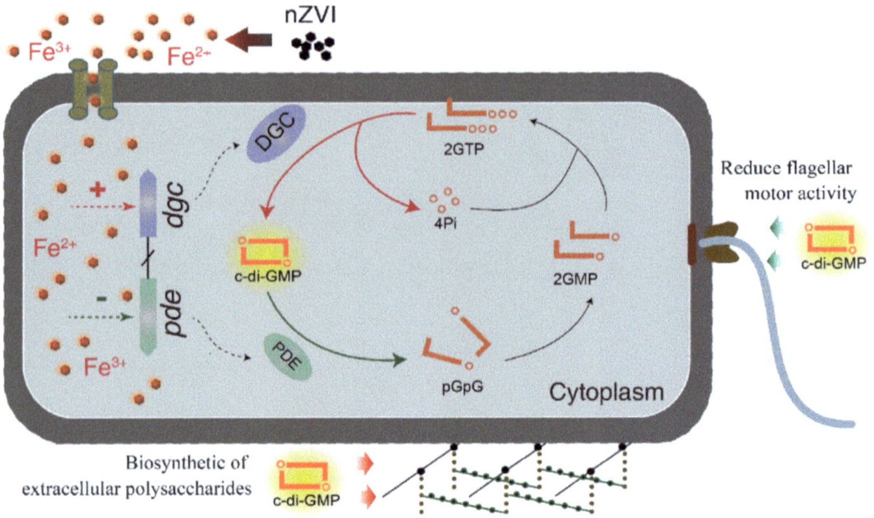

Fig. 2 Possible mechanism of nZVI promoting anammox. The addition of nZVI significantly increased the abundance of the DGC gene and decrease the abundance of the PDE gene, which resulted in the enrichment of c-di-GMP in microbial cells. The enrichment of c-di-GMP reduced the motility of microorganisms in the reactor and promoted the secretion of extracellular polymers by bacteria, which was beneficial to the formation of sludge particles in the anammox reactor. In addition, c-di-GMP participated in the hydrolysis of the inhibitory protein of bacterial division, thereby accelerating the division of anammox bacteria and increasing the abundance of anammox bacteria. Reprinted with permission from Wang, Z., Liu, X., Ni, S. Q., Zhuang, X., Lee, T., 2021. Nano zero-valent iron improves anammox activity by promoting the activity of quorum sensing system. Water Res. 202, 117,491. Copyright © 2021 Elsevier Ltd.

cells in the reactor which promoted the secretion of bacterial extracellular polymers that further facilitated granule formation in the anammox reactor.

The nitrogen removal ability of anammox bacteria was also enhanced in the study conducted by Erdim et al. (2019) using nZVI. A sequencing batch reactor (SBR) was used for this study in a continuous mode with an enriched anammox culture containing *Candidatus Brocadia anammoxidans* and *Candidatus Scalindua.* Aqueous phase reduction of ferrous ions was carried out for the synthesis of nZVI using $NaBH_4$ at room temperature. Thereafter varying concentrations of nZVI ranging from 0.04 to 5000 parts per billion (ppb) were fed in the SBR which resulted in a 58% and 73% increase in the ammonium and nitrite removal rates, respectively. The SAA was increased in various time periods of the reactor operation. However, the overall SAA measurements showed minimal difference in the presence of nZVI which was attributed to the formation of gas pockets inside the anammox granules. TEM images further confirmed this hypothesis wherein an hourly increase in the SAA values could be due to the sudden burst of N_2 gas inside the cavity of the granules. The concentration of EPS was reduced with a concomitant increase in the nZVI content. A steady rise in EPS content was attained in the presence of 5000 ppb of nZVI which highlighted enhanced extracellular microbial protein secretion in the presence of nZVI. Quantitative real-time polymerase chain reaction (qPCR) results indicated improvement in the population of anammox bacteria in the presence of nZVI. Furthermore, high-resolution melting (HRM) analysis revealed the formation of four different anammox clades in the reactor.

3 Metal Oxide Nanoparticles

Various nanostructured metal oxides may have a pronounced effect on the growth and metabolism of anammox bacteria, which is discussed in detail in this section. Zhang et al. (2018a) reported a comparative study of toxicity imparted by CuNPs and copper oxide nanoparticles (CuONPs) wherein commercially manufactured NPs with an average size of 10–30 nm and 40 nm were used, respectively. The initial NRE and NRR of the reactors containing mature anammox granules were 88.2% and 12.3 kg of N/m^3/day, respectively. After 30 days of the reactor operation, 5 mg/L of both CuNPs and CuONPs was introduced into the system separately which released 3.21 $\pm$ 0.46 and 0.33 $\pm$ 0.11 mg/L of Cu(II) ions in the influent. The SAA of granules was slightly enhanced by 10.5% in the presence of 1 mg of Cu/L CuNPs while it was inhibited by 90.9% when the concentration was increased to 5 mg of Cu/L CuNPs. On the contrary, the addition of 5 mg of Cu/L CuONPs only resulted in 12.8% inhibition of SAA activity. Surprisingly, this inhibitory effect of CuONPs was attenuated by increasing the concentration of NPs to 40 mg or 160 mg of Cu/L. Such improvement in tolerating high concentrations of CuONPs was attributed to the long-term acclimatization of the anammox consortia along with aggregation of the CuONPs which alleviated its toxic effect. The abundance of *HzsA* gene was also observed in the presence of CuONPs at a concentration of 40–160 mg of Cu/L. The protein content increased sharply with a subsequent increase in the CuONPs concentration while the EPS content was 315.0 $\pm$ 15.3 mg/g of VSS in the presence of 5 mg Cu/L CuONPs. Redundancy analysis then confirmed the minimal toxic effect of CuONPs on the environmental factors of the anammox sludge.

Li et al. (2018b) also investigated the long-term exposure effect of iron oxide nanoparticles (Fe_3O_4NPs) on the anammox process of constructed wetlands. Three unplanted subsurface-flow constructed wetlands (USFCWs) were used in this study with a stable nitrogen loading rate (NLR), NRR, and HRT of 58 g of N/(m^3 day), 44 g of N/(m^3 day), and 48 h, respectively. The total nitrogen removal was improved with a removal efficiency of 77.5% after 23 days of operation in the presence of Fe_3O_4NPs along with SAA of 0.026 kg-N/(kg VSS day). Moreover, the Fe^{3+} and Fe^{2+} ions that were released from the Fe_3O_4NPs were suggested to improve the anammox activity as well. The abundance of certain anammox bacterial species such as *Candidatus Anammoxoglobus* and *Parcubacteria_genera_incertae_sedis* was increased to 15.2 and 27.7%, respectively, in the presence of 10 mg/L of Fe_3O_4NPs. Thereafter, the prediction of functional genes was carried out using the phylogenetic investigation of communities by reconstruction of unobserved states (PICRUSt) wherein genes responsible for the transport of inorganic ions and lipid metabolism were upregulated from 4.85 and 3.1 to 5.25 and 3.5%, respectively. Moreover, the level of extracellular lactate dehydrogenase (LDH) was not significantly increased in the presence of 10 mg/L of Fe_3O_4NPs which highlighted the maintenance of stable cell membrane integrity even after Fe_3O_4NPs exposure. On the contrary, reactive oxygen species (ROS) production was significantly increased by 748% inside the cells after exposure to 10 mg/L of Fe_3O_4NPs which suggested an increase in the intracellular oxidative stress facilitated by the Fe_3O_4NPs. Levels of certain enzymes such as nitrite reductase (NIR) and hydrazine dehydrogenase (HDH) were decreased in the presence of 10 mg/L of Fe_3O_4NPs as well.

Xu et al. (2020) also investigated the long-term effects of Fe_3O_4NPs on the anammox process. A UASB reactor was inoculated with anammox granular sludge having an initial sludge concentration of 18.3 g of VSS/L along with continuous pumping of synthetic wastewater. Moreover, the concentration of both ammonium and nitrite was maintained at 280 mg/L with an HRT of 0.9 h. The concentration of Fe_3O_4NPs was gradually increased from 0 mg/L to 200 mg/L after 145 days of reactor operation. There was no significant alteration in the NRR and NRE after Fe_3O_4NPs supplementation. However, the SAA of anammox sludge increased from 239.0 ± 21.5 mg of total nitrogen/g of VSS/day to 307.0 ± 13.8 mg of total nitrogen/g of VSS/day in the presence of 200 mg/L of Fe_3O_4NPs. The concentrations of VSS and suspended solids (SS) also increased from 21.6 and 36.9 g/L to 23–36 and 38–43 mg/L, respectively, after the addition of Fe_3O_4NPs. Likewise, the heme c content also increased from 1.59 ± 0.01 to 2.67 ± 0.05 μmol/g of VSS in the presence of 200 mg/L of Fe_3O_4NPs which suggested the enhanced metabolic state of the anammox bacteria using the Fe(II) and Fe(III) ions released from the Fe_3O_4NPs. The concentration of EPS also gradually increased by 72.7% with the subsequent addition of 200 mg/L of Fe_3O_4NPs which was speculated to prevent the accumulation of NPs inside the anammox bacteria. Thereafter, 16S rRNA gene sequencing demonstrated the dominance of *Candidatus Kuenenia* throughout the entire reactor operation which was increased by 32.9% in the presence of 200 mg/L of Fe_3O_4NPs. Other bacterial genera such as *Candidatus Hydrogenedens, Armatimonadetes gp5, Caldilinea, Ignavibacterium, Chiayiivirga, Gp10, Phaselicystis, Gp6, Hyphomicrobium, Nitrosomonas, Oceanibaculum, Gaiella, Litorilinea, Phycisphaera, Gp16, Ornatilinea, Mesorhizobium,* and *Conexibacter* were also present in the reactor. Therefore, the study demonstrated the long-term effect of Fe_3O_4NPs on the reactor operation which positively impacted the anammox process.

Ma et al. (2022) recently demonstrated the toxicity of alleviated level of magnesium oxide nanoparticles (MgONPs) against anammox granular sludge through the addition of exogenous proteins. Commercial MgONPs having a purity and average size of 99% and 50 nm, respectively were added to a laboratory-scale UASB reactor containing anammox granular sludge that had more than 20% abundance of *Candidatus Kuenenia stuttgartiensis*. The SAA of anammox sludge was reduced by 80% in the presence of 200 mg/L of MgONPs within 4 h of reactor operation. Moreover, the IC_{50} of MgONPs for inhibition of anammox sludge was 69.5 mg/L. The NRE of the sludge was maintained at around 91.0 ± 6.0% in the presence of low concentrations (2 and 5 mg/L) of MgONPs however, at 50 mg/L of MgONPs, the NRE was drastically reduced to 26.0 ± 2.0% which resulted in higher concentrations of ammonia and nitrite in the effluent. The microbial community was also altered after the introduction of 50 mg/L of MgONPs with an increase in the abundance of *Denitratisoma* from 4.3 ± 1.2% to 11.9 ± 2.3% which promoted denitrification process. Thereafter, 5 mg/L of bovine serum albumin (BSA) was supplemented in the sludge which increased the SAA from 361.4 ± 52.7 to 470.6 ± 34.7 mg of total nitrogen/ g of VSS/day in the presence of 50 mg/L of MgONPs. Around 97% of anammox activity was recovered after the addition of BSA. Multiple spectral analyses such as ultraviolet–visible (UV–vis), Fourier transform infrared (FTIR) as well as fluorescence spectra confirmed the adsorption of MgONPs by BSA which was the major mechanism of alleviation of NPs stress in the anammox sludge. TEM micrographs also displayed the formation of an eco-corona on the surface of MgONPs which also facilitated reduced biotoxicity of the MgONPs.

Maghemite nanoparticles (MHNPs) were also demonstrated to positively impact the anammox abundance in a high-rate anammox reactor (Zhang et al., 2018b). The anammox seeding sludge was obtained from lab-scale UASB reactors. The addition of 200 mg/L of MHNPs showed no adverse effect on the anammox activity as well as maintained the cell membrane integrity. Moreover, the concentrations of NH_4^+ and NO_2^- were reduced to 60.6 ± 2.9 and 7.8 ± 3.1 mg of N/L, respectively after the concentration of MHNPs was gradually increased from 1 to 200 mg/L. In addition, the specific anammox activity (SAA) was significantly enhanced from 340.1 ± 25.6 mg of total nitrogen/g of volatile suspended solids (VSS)/day to 478 ± 26.7 mg of total nitrogen/ g of VSS/day in the presence of 200 mg/L of MHNPs. Further, the MHNPs exhibited a maximum release of 2.64 ± 0.16 mg/L of Fe(III) ions that showed no inhibitory effect on the anammox process. The microbial community was abundant with *Candidatus Kuenenia* as indicated through high-throughput sequencing. The EPS content also increased to 272.6 ± 16.7 mg/g of VSS after supplementation of 200 mg/L of MHNPs. Hence, the specific biocompatibility of MHNPs for increasing the anammox activity was evident in this study.

Xu et al. (2019) investigated the performance of anammox bacteria in the presence of manganese oxide nanoparticles (MnO_2NPs) using a UASB reactor. MnO_2NPs were introduced after 25 days of the reactor operation containing anammox seeding granules in the inorganic synthetic wastewater. The addition of 200 mg/L of MnO_2NPs demonstrated a positive impact on the nitrogen removal ability with a maximum NRE of 93.1% along with the highest SAA of 657.3 ± 9.3 mg of total nitrogen/g VSS/day which was 1.6-folds higher than the control reactor in the absence of MnO_2NPs. The nitrogen loading rate (NLR) was thus, also enhanced from 13.8 ± 0.27 to 14.2 ± 0.33 kg N/m^3/day because of increased SAA. The EPS content in terms of the amount of the total concentration of polysaccharide and protein was also increased by 1.3-fold to a maximum of 481.5 ± 13.4 mg/g of VSS in the presence of 200 mg/L of MnO_2NPs. Analysis of the microbial community in the presence of MnO_2NPs using high-throughput sequencing revealed an increase in the abundance of *Candidatus Kuenenia* from 17.3% to 23.9% after the addition of 200 mg/L of MnO_2NPs which facilitated improved anammox process and enhanced nitrogen removal ability as well.

Titanium dioxide nanoparticles (TiO_2NPs) were also demonstrated to exhibit short-term inhibitory effects on anammox activity (Tuğba et al., 2020). The size of the commercial TiO_2NPs was less than 25 nm which was suspended in 0.1 mM of SDBS. Acute exposure to 300 mg/L of TiO_2NPs for 24 h resulted in 80.57 ± 1.17% inhibition in the nitrogen removal ability of anammox bacteria. Inhibition model studies then revealed a best-fit curve obtained using the non-linear second-order polynomial (quadratic) model which provided an IC_{50} value of 154 mg/L. Scanning electron microscopy (SEM) images then demonstrated the formation of TiO_2NPs clusters which can adsorb to the surface of bacteria and alter the substrate transfer rate of the bacteria from the environment which in turn reduced the NRR. The total EPS content was reduced by 92.19% in the presence of 300 mg/L of TiO_2NPs, which highlighted the adsorption of TiO_2NPs by the EPS.

Zhang et al. (2018c) also investigated the impact of zinc oxide nanoparticles (ZnONPs) on the anammox process in wastewater treatment. Commercially produced ZnONPs were used in this study with an average size of 30 ± 10 nm and dispersed in distilled water containing 0.1 mM of sodium dodecylbenzene sulfonate (SDBS) that acted as the dispersing agent. Mature anammox granules were inoculated in a UASB reactor and cultivated for 30 days in the presence of synthetic wastewater composed of substrates, minerals, and trace elements. Thereafter, varying amounts of ZnONPs were added to the influent. No visible fluctuations were observed in the concentrations of NH_4^+, NO_2^-, and NO_3^- in the presence of 1.0–5.0 mg/L of ZnONPs. However, the NRE was decreased by 11.3% in the presence of 10.0 mg/L of ZnONPs after 60 days of the reactor operation along with an increase in the effluent NH_4^+–N and NO_2^- concentrations to 259.2 and 237.5 mg of N/L, respectively. This ZnONPs shock was alleviated with immediate withdrawal of the NPs resulting in a gradual decrease in the effluent NO_2^- concentration. Almost 90% of the nitrogen removal capacity was lost due to 10.0 mg/L of ZnONPs shock for 3 days which was mainly attributed to the accumulation of toxic Zn(I) ions in the biomass of anammox bacteria that altered the reactor performance. A subtle response was observed after another ZnONPs shock which also highlighted the resilience and the transitory inhibition. The SAA of the anammox biomass after fifth shock was 309.2 ± 53.3 mg of total nitrogen/g of VSS/day. The relative abundance of *Hzs*A gene that acted as the specific marker for anammox bacteria was reduced during the shock period which was then rebounded during the recovery phase. In addition, 16S rRNA gene sequencing analysis of the microbial community in the reactor revealed the highest abundance of *Candidatus Kuenenia,* which remained stable after five cycles of ZnONPs "shock-recovery" process.

Song et al. (2018) also evaluated the short-term interaction of ZnONPs with anammox granules and investigated its implication on the nitrogen removal ability of anammox bacteria. Commercial ZnONPs with an average size of 30 ± 10 nm were used in this study along with laboratory-scale anammox granules, which had the highest abundance of *Candidatus Brocadia.* A severe reduction in SAA to 2.5 mg of N g/VSS was observed in the presence of 20 mg of N g/VSS of ZnONPs within

3 h of operation. The anammox activity was completely inhibited when ZnONPs concentration was further increased. The half inhibitory concentration (IC_{50}) value of ZnONPs was 11.7 ± 0.9 mg g/VSS. Energy-dispersive spectroscopy (EDS) results then displayed the distribution of Zn throughout the surface of the anammox granules along with up to 40% of Zn in the EPS produced by the anammox bacteria. A non-competitive inhibition on the nitrite conversion was imposed by the ZnONPs as per the Haldane kinetic model simulations which resulted in a significant decrease in nitrogen removal rate whereas ammonium conversion kinetics showed competitive inhibition that exhibited a sixfold increase in the affinity constant. Hence, the formation of colloidal suspension of ZnONPs followed by the release of Zn(II) ions was proposed as the primary mechanism of inhibition of the anammox process.

In another study, Zhao et al. (2019) reported the effects of ZnONPs on the anammox granules at ambient temperature. Commercial ZnONPs in the form of a powder with an average size less than 100 nm were used in this experiment wherein SDBS was added as a dispersing agent. The anammox activity was reduced by 64.0% as compared to the control after the addition of 150 mg/L of ZnONPs. The total protein and polysaccharide contents also decreased from 69.8 and 23.2 mg/g of VSS to 37.9 and 9.8 mg/g of VSS in the presence of 150 mg/L of ZnONPs, respectively. Moreover, EPS was produced by the anammox bacteria in response to the stress exerted by the NPs which adsorbed 5.16 mg/L of ZnONPs and alleviated the NPs' stress. Hence, EPS produced by the anammox biomass was attributed to the protection of the granules through binding the Zn(II) cations that were released from the ZnONPs. The amount of LDH released by the anammox biomass was also increased after 24 h of ZnONPs exposure along with increased production of ROS that indicated an alteration in the cell membrane integrity as well as induction of oxidative stress, respectively. Although dissolution of Zn(II) cations was proposed as the primary mechanism of anammox inhibition, undissolved ZnONPs contributed to the inhibition with increasing concentrations of the NPs. About 28.4% inhibition was observed at a concentration of 150 mg/L whereas only 3.6% inhibition was contributed by Zn(II) cations. Additionally, the heme and ATP content also decreased with a concomitant increase in the ZnONPs concentration.

4 Carbon-Based Nanomaterials

Metabolic alteration in the anammox bacteria can even be brought about by the carbon-based nanomaterials. Elreedy et al. (2021) reported the stimulation of anammox granular sludge using graphene nanosheets (GNs) and γ-Fe_2O_3 nanoparticles (NPs). Consortia of anammox bacteria were obtained from a lab-scale up-flow anaerobic sludge blanket (UASB) reactor that was fed with 124.0 ± 5.6 mg/L and 148.2 ± 3.1 mg/L of $NH_4{}^+$–N and $NO_2{}^-$–N, respectively, at a hydraulic retention time (HRT) of 14 h. The nanomaterials were selected due to the conductivity and electron shuttling of GNs and γ-Fe_2O_3NPs, respectively. A total nitrogen removal rate (NRR) of 46 ± 3.1% along with the removal of 86.5 ± 2.7% and 97.1 ± 0.5% of $NH_4{}^+$–N and

NO_2^-–N was effectively removed in the presence of 10 mg/L of GNs, respectively, along with 1.1-fold augmentation of hydrazine dehydrogenase enzyme activity as compared to the control. The underlying mechanism is schematically represented in Fig. 3. The granulation process of anammox bacteria was upregulated in the presence of 10 mg/L of GNs, which was attributed to the enhanced hydrophobic interaction of GNs with the extracellular polymeric substances (EPS) that were produced by the anammox bacteria. Likewise, 100 mg/L of γ-Fe_2O_3NPs enhanced the NRR by 55 $\pm$ 3.8%, however, this increase in the removal rate was attributed to the adsorption efficiency of the NPs. Thereafter, 16S rRNA gene sequencing results revealed the 12.3% abundance of *Candidatus Jettenia* in the presence of 10 mg/L of GNs, which was reduced to 8.3% when 100 mg/L of γ-Fe_2O_3NPs were present. Therefore, the addition of nanosheets positively impacted the anammox process in this study.

Similarly, Wang et al. (2013) also reported the improvement of anammox activity in the presence of GO in a dose-dependent manner. Modified Hummers method was carried out for the preparation of GO using natural graphite powder after which TEM analysis was conducted for evaluation of its microstructure which showed the presence of distinct GO nanofilms with a wrinkled structure composed of multiple GO layers which were stacked and displayed interlaced edges. The X-ray diffraction

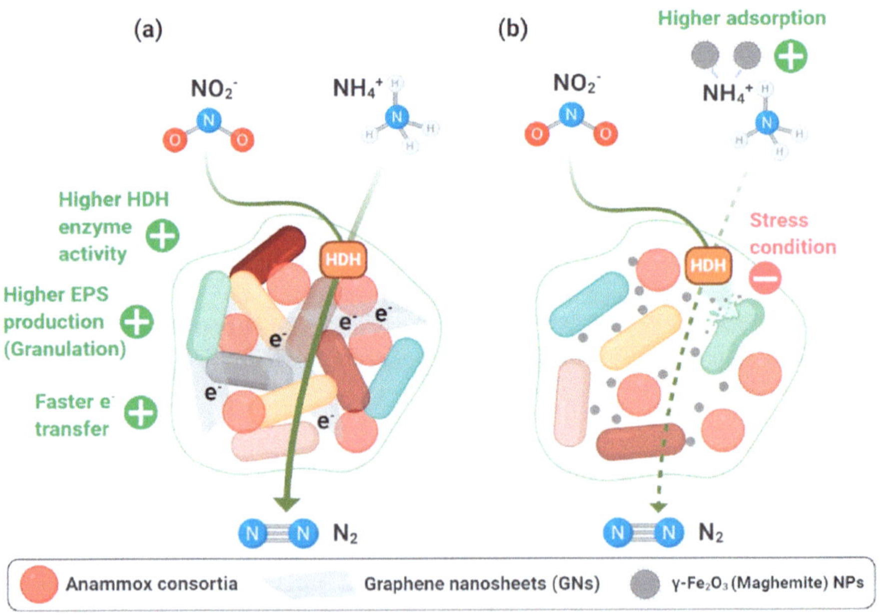

Fig. 3 Proposed enhancement mechanism of ammonium removal via anammox consortia in case of supplementing graphene nanosheets (GNs) **a** and γ-Fe_2O_3 (maghemite) nanoparticles (NPs) **b**. EPS, extracellular polymeric substances; HDH, hydrazine dehydrogenase. Reprinted with permission from Elreedy, A., Ismail, S., Ali, M., Ni, S. Q., Fujii, M., Elsamadony, M., 2021. Unraveling the capability of graphene nanosheets and γ-Fe_2O_3 nanoparticles to stimulate anammox granular sludge. J. Environ. Manage. 277, 111,495. Copyright © 2020 Elsevier Ltd.

(XRD) pattern then demonstrated a sharp diffraction band at 26.4° corresponding to the graphite plane. Thereafter, the anammox activity was increased by 10.26% in the presence of 0.1 g/L of GO, which also contributed to a maximum polysaccharide, protein, and total EPS content of 42.5, 125.7, and 168.2 mg/g of VSS, respectively. The TEM micrographs revealed an efficient cell attachment region that was provided by GO which in turn, enhanced the anammox activity.

Tomaszewski et al. (2019) also reported the long-term effect of reduced graphene oxide (rGO) on the anammox process at low temperatures. The experiments were carried out in an SBR suspended with anammox biomass and mineral media followed by the addition of commercially available rGO. TEM micrographs of the rGO after 109 days of incubation in the anammox reactor at a low-temperature range of 10–15 °C displayed its degradation, which was further confirmed by electron energy loss spectroscopy (EELS) and Raman spectral analysis that showed deposition of oxygen and calcium on the surface. The rGO stimulated the anammox activity by 17% wherein the bacterial growth was enhanced at higher temperatures while the anammox reaction rate was improved at temperature less than 15 °C.

5 Conclusion and Future Perspectives

Several nanoparticles have shown notable effect on the anammox process and hence the exploration of the underlying mechanism has gained tremendous attention more recently. The enzymatic modulation of the bacterial metabolism in the presence of the metal, metal oxide, and the carbon-based nanoparticles might play a critical role in the overall nitrogen removal. The molecular effect in the level of transcription and translation results in upregulation of certain genes while downregulating others. Hence, employing the integrated knowledge of metagenomics, proteomics, and metabolomics will help in the better understanding of the underlying mechanism.

It is important to note that certain nanoparticles are reported to adversely affect the overall efficiency of the anammox process that might be attributed to the toxicity of the nanoparticles. The chemical and/or physical synthesis process for fabrication of nanoparticles often use hazardous reaction conditions and toxic reducing as well as stabilizing agents that render the resulting nanoparticles toxic. Hence, biologically synthesis using bacteria, fungi, algae and plants are considered as environmentally benign green route for the synthesis of nanoparticles which are more biocompatible. Hence, such biogenic nanoparticles should be included in the studies for exploring their role in the augmentation of the anammox process.

Additionally, anammox-related enzymes such as nitrite oxidoreductase (NirS), hydrazine dehydrogenase (Hdh), and hydrazine synthase (HZS) can be immobilized in natural polymers such as alginate, chitosan and gelatin for stability, recovery and recycle. Mesoporous nanoparticles like silica and carbon nanotubes (CNTs) can be used for immobilization of such enzymes for enhanced anammox activity. This will be rapid, efficient and can effectively bypass the biomass generation step. In view of the background integration of nanotechnology with anammox-based wastewater treatment process can serve as a powerful strategy to ensure clean environment.

Acknowledgements Dr. Sougata Ghosh acknowledges the Program of National Postdoctoral and Postgraduate System approved by PMU-B Board Committees (Contract No. B13F660065), Thailand (Ref. No. 6202/1592 dated May 01, 2023) and Kasetsart University, Bangkok, Thailand for Post Doctoral Fellowship and funding under Reinventing University Program (Ref.No.6501.0207/9219 dated 14th September, 2022).

References

Bhagwat, T. R., Joshi, K. A., Parihar, V. S., Asok, A., Bellare, J., & Ghosh, S. (2018). Biogenic copper nanoparticles from medicinal plants as novel antidiabetic nanomedicine. *World Journal of Pharmaceutical Research, 7*(4), 183–196.

Bloch, K., Pardesi, K., Satriano, C., & Ghosh, S. (2021). Bacteriogenic platinum nanoparticles for application in nanomedicine. *Frontiers in Chemistry, 9*, 624344.

Cheng, Y. F., Li, G. F., Ma, W. J., Xue, Y., Liu, Q., Zhang, Z. Z., & Jin, R. C. (2020). Resistance of anammox granular sludge to copper nanoparticles and oxytetracycline and restoration of performance. *Bioresource Technology, 307*, 123264.

Elreedy, A., Ismail, S., Ali, M., Ni, S. Q., Fujii, M., & Elsamadony, M. (2021). Unraveling the capability of graphene nanosheets and γ-Fe_2O_3 nanoparticles to stimulate anammox granular sludge. *Journal of Environmental Management, 277*, 111495.

Erdim, E., Özkan, Z. Y., Kurt, H., & Kocamemi, B. A. (2019). Overcoming challenges in mainstream anammox applications: Utilization of nanoscale zero valent iron (nZVI). *Science of the Total Environment, 651*, 3023–3033.

Li, G. F., Huang, B. C., Zhang, Z. Z., Cheng, Y. F., Fan, N. S., & Jin, R. C. (2019). Recent advances regarding the impacts of engineered nanomaterials on the anaerobic ammonium oxidation process: Performances and mechanisms. *Environmental Science Nano, 6*, 3501.

Li, J., Li, J., Gao, R., Wang, M., Yang, L., Wang, X., Zhang, L., & Peng, Y. (2018a). A critical review of one-stage anammox processes for treating industrial wastewater: optimization strategies based on key functional microorganisms. *Bioresource Technology, 265*, 498–505.

Li, H., Chi, Z., & Yan, B. (2018b). Insight into the impact of Fe_3O_4 nanoparticles on anammox process of subsurface-flow constructed wetlands under long-term exposure. *Environmental Science and Pollution Research, 25*, 29584–29592.

Ma, W. J., Wang, X., Zhang, J. T., Guo, J. Y., Lin, Y. X., Yao, Y. X., Li, G. F., Cheng, Y. F., Fan, N. S., & Jin, R. C. (2022). Adding exogenous protein relieves the toxicity of nanoparticles to anammox granular sludge by adsorption and the formation of eco-coronas. *Environmental Science: Nano.*

Peng, M. W., Yu, X. L., Guan, Y., Liu, P., Yan, P., Fang, F., Guo, J., & Chen, Y. P. (2019). Underlying promotion mechanism of high concentration of silver nanoparticles on anammox process. *ACS Nano, 13*, 14500–14510.

Shende, S., Joshi, K. A., Kulkarni, A. S., Shinde, V. S., Parihar, V. S., Kitture, R., Banerjee, K., Kamble, N., Bellare, J., & Ghosh, S. (2017). *Litchi chinensis* peel: A novel source for synthesis of gold and silver nanocatalysts. *Global Journal of Nanomedicine, 3*(1), 555603.

Song, Y. X., Chai, L. Y., Tang, C. J., Xiao, R., Li, B. R., Wu, D., & Min, X. B. (2018). Influence of ZnO nanoparticles on anammox granules: The inhibition kinetics and mechanism analysis by batch assays. *Biochemical Engineering Journal, 133*, 122–129.

Tomaszewski, M., Cema, G., Ciesielski, S., Łukowiec, D., & Ziembińska-Buczyńska, A. (2019). Cold anammox process and reduced graphene oxide-Varieties of effects during long-term interaction. *Water Research, 156*, 71–81.

Tuğba S. A. R. I., Yar-Bektaş, A. Ö., Akgül, D. (2020). Short-term inhibitory effects of TiO_2 NPs on anammox process. *Biotech Studies, 29*(1), 29–37

Wang, D., Wang, G., Zhang, G., Xu, X., & Yang, F. (2013). Using graphene oxide to enhance the activity of anammox bacteria for nitrogen removal. *Bioresource Technology, 131*, 527–530.

Wang, Z., Liu, X., Ni, S. Q., Zhuang, X., & Lee, T. (2021). Nano zero-valent iron improves anammox activity by promoting the activity of quorum sensing system. *Water Research, 202*, 117491.

Xu, J. J., Cheng, Y. F., Zhu, X. L., Zhu, W. Q., & Jin, R. C. (2019). The performance and microbial community in response to MnO_2 nanoparticles in anammox granular sludge. *Chemosphere, 233*, 625–632.

Xu, J. J., Cheng, Y. F., & Jin, R. C. (2020). Long-term effects of Fe_3O_4 NPs on the granule-based anaerobic ammonium oxidation process: Performance, sludge characteristics and microbial community. *Journal of Hazardous Materials, 398*, 122965.

Zhang, Z. Z., Cheng, Y. F., Wu, J., Bai, Y. H., Xu, J. J., Shi, Z. J., & Jin, R. C. (2018a). Discrepant effects of metal and metal oxide nanoparticles on anammox sludge properties: A comparison between Cu and CuO nanoparticles. *Bioresource Technology, 266*, 507–515.

Zhang, Z. Z., Cheng, Y. F., Bai, Y. H., Xu, J. J., Shi, Z. J., Zhang, Q. Q., & Jin, R. C. (2018b). Enhanced effects of maghemite nanoparticles on the flocculent sludge wasted from a high-rate anammox reactor: Performance, microbial community and sludge characteristics. *Bioresource Technology, 250*, 265–272.

Zhang, Z. Z., Cheng, Y. F., Bai, Y. H., Xu, J. J., Shi, Z. J., Zhang, Q. Q., & Jin, R. C. (2018c). Transient disturbance of engineered ZnO nanoparticles enhances the resistance and resilience of anammox process in wastewater treatment. *Science of the Total Environment*, 622–623, 402–409.

Zhao, J., Zhang, B., & Zuo, J. (2019). Response of anammox granules to ZnO nanoparticles at ambient temperature. *Environmental Technology and Innovation, 13*, 146–152.

Anammox Cell Biology, Metabolism, Growth, and Genetics

Sonia Sethi

Abstract One of the most recent contributions to the biogeochemical nitrogen cycle is anaerobic ammonium-oxidizing (anammox) bacteria. Anaerobic ammonium-oxidizing bacteria (anammox) obtain their energy for development by converting ammonium and nitrite into dinitrogen gas in the absence of oxygen. The species were thought to be extinct until around 15 years ago due to the believed inert nature of ammonium in anoxic circumstances. These slow-growing microorganisms are members of the Brocadiales order, which is related to the Planctomycetes. Membrane systems encircling the distinct cellular compartments are formed of unique "ladderane" lipid molecules. Anammox bacteria have a compartmentalized cell design with a central cell compartment, the "anammoxosome." The intermediate production of hydrazine, a highly reactive and poisonous molecule, appears to be required for nitrogen generation. A metagenomics technique was used to assemble the genome of the anammox bacterium *Kuenenia stuttgartiensis* from a complex microbial community cultivated in a sequencing batch reactor (74% enriched in this bacterium). The anammox metabolism was in silico reconstructed using the assembled genome, and genes most likely involved in the process were identified. Anammox bacteria have been found in a variety of oxygen-depleted marine and freshwater environments, including oceans, seas, estuaries, marshes, rivers, and big lakes. Anammox bacteria may create more than half of the N_2 gas discharged in the maritime environment. For the elimination of ammonia nitrogen, the anammox process is an appealing alternative to conventional wastewater treatment technologies.

Keywords Nitrogen cycle · Anaerobic ammonium oxidation · Metagenome · Anammoxosome · Ladderane lipids · Brocadiales

S. Sethi (✉)
Dr. B. Lal Institute of Biotechnology, Malviya Industrial Area, Malviya Nagar, Jaipur, India
e-mail: soniakaura@drblal.com

M. P. Shah (ed.), *Anammox Technology in Industrial Wastewater Treatment*,
https://doi.org/10.1007/978-981-99-3459-1_4

1 Introduction

Dinitrogen gas makes up 78% of our atmosphere. N_2 is continually taken up and created at a turnover rate of one billion years, despite its relatively stable concentration across geological eras. The nitrogen cycle (N-cycle), which is mostly driven by microbes, balances nitrogen conversion through nitrogen fixation, anammox, denitrification, and other processes. Human actions, on the other hand, have the potential to disrupt the equilibrium not only locally but also globally. The Haber Bosch technique for industrial nitrogen fixation has been critical for modern agriculture. Fixed nitrogen (ammonium, nitrate) has a number of disadvantages, including fish poisoning, eutrophication of freshwater habitats, emission of the strong greenhouse gas nitrous oxide (N_2O), and acid rain.

The horizontal and vertical spread of oxygen-depleted zones in the seas is caused by fertilizer run-off and the drainage of vast volumes of organic waste into rivers and estuaries, with side effects ranging from fish habitat constriction to increased N_2O production (Stramma et al., 2012). Due to our poor understanding of microbial N-cycle mechanisms, the impacts are only partially understood. Due to the activities of neglected essential actors, disregarded partial processes, and very complicated microbial interactions, the cycle that was considered to be pretty well-known for more than a century appears to be far more intricate. Regardless of the current level of knowledge, wastewater treatment systems are necessary to prevent the release of excessive amounts of fixed nitrogen into the environment, particularly in urban areas.

2 Physical Properties of Inorganic Nitrogen Compounds

Nitrogen (N) belongs to group 5B and has oxidation states ranging from 3 to 5. The nitrogen atom unites with hydrogen, oxygen, or other nitrogen atoms in each oxidation state. In this approach, each oxidation state has at least one distinct inorganic molecule. Because the oxidation state of N in a particular environment is governed by kinetics (the activation energy of the N-compounds is large), rather than thermodynamic equilibrium, all oxidation states are feasible in aqueous systems. The majority of nitrogen on the planet is in solid form (rocks). However, the most essential nitrogen source available to organisms is the dinitrogen gas in the atmosphere (79% vol.).

3 NH^{4+} and the Environment

All living creatures require NH^{4+}, which is typically a growth-limiting nutrient. Environmental risks such as eutrophication and global warming are caused by excessive deposits in the environment in densely inhabited places and the excessive use of fertilizers in contemporary agriculture (Vitousek et al., 2009; Canfield et al., 2010). This volume is around 3.5 times greater than the estimated acceptable threshold for nitrogen deposition (35 million tons per year) and has been steadily increasing (Rockström et al., 2009). However, microbial activity may remove NH^{4+} and bacteria that can convert NH^{4+} to other fixed nitrogen compounds have been recognized for over a century.

In the presence of oxygen, aerobic ammonium-oxidizing bacteria (AOB) and archaea (AOA) may convert NH^{4+} to NO_2, which is then converted to nitrate (NO_3) by nitrite-oxidizing bacteria (Monteiro et al., 2014; Daims et al., 2016). Nitrification is the name for these oxidative reactions. NO_2 and NO_3 can be reduced to N_2 by denitrifiers or to NH^{4+} by dissimilatory nitrite/nitrate-reducing (DNRA) bacteria in the absence of oxygen (Lu et al., 2014). The opportunities provided by nitrifiers and denitrifiers are used in traditional wastewater treatment to remove excess fixed nitrogen from wastewater streams. Every other N_2 molecule in the atmosphere is now created by a group of microorganisms that were previously thought to be non-existent until roughly 15 years ago: anaerobic bacteria.

4 Anammox Bacteria

Because ammonium is a rather inert molecule, AOB and AOA rely on oxygen's oxidative capacity to convert it to the more accessible hydroxylamine (NH_2OH) (Arp et al., 2002). The search for microbial activities that convert ammonium without oxygen may have been impeded by this perceived tight reliance on oxygen (anoxically). Such procedures were previously thought to be impossible. Even yet, the key discovery that ammonium was seemingly depleted in a pilot-scale bioreactor operated under anoxic conditions was made 20 years ago. This discovery sparked a quest for the microbes to blame. These slow-growing critters, known as anaerobic anammox bacteria, might be enhanced for further research using specialized procedures (Kartal et al., 2010). More than 10 species and subspecies of anammox bacteria have been recorded in situations where fixed nitrogen is eliminated in the absence of oxygen since their first identification, which permitted the development of molecular methods for their detection (Song & Tobias, 2011; Oshiki et al., 2015).

Anammox bacteria are thought to be responsible for 30–70% of the yearly emission of N_2 into the atmosphere (Devol, 2015). Furthermore, these microbes are currently being used as a low-cost, ecologically friendly alternative to traditional wastewater treatment (Kartal et al., 2010). Anammox bacteria are remarkable creatures in a variety of ways. They are members of the Eubacterial Phylum of Planctomycetes, which are distinguished by their complex cell plan (Fuerst & Sagulenko, 2013). This is also true for anammox bacteria, which have a huge vacuolar cell organelle called the anammoxosome that contains the proteins involved in central catabolism.

Anammox bacteria produce nitrogen by combining the reactions of ammonium oxidation and nitrite reduction. Because its cells could fix CO_2, Anammox was also a chemolithoautotroph. In addition to its unique metabolic characteristics, the anammox bacteria may be well-suited for usage as an efficient, cost-effective, and environmentally friendly nitrogen removal alternative to typical wastewater treatment. Anammox bacteria are extraordinary in a number of ways. A unique catabolic mechanism is required for their ability to anaerobically oxidise ammonium. According to 16S ribosomal ribonucleic acid (16S rRNA) study (Fuerst & Sagulenko, 2011), the organisms belong to the Planctomycetes superphylum, which includes *Verrucomicrobia and Chlamydiae.*

Nine anammox species have been isolated thus far, however, none of them in pure culture, earning them Candidatus rank. The 16S rRNA gene has a sequence identity between 87 and 99%. The deposited 16S rRNA gene sequences encompass a broad range, allowing for the emergence of a slew of new species, subspecies, and strains, each with its own niche in the vast array of environments where the organisms live. The known species are split into five genera, all of which belong to the same order, *Brocadiales,* which is a monophyletic clade within the Planctomycetes (Fuerst & Sagulenko, 2011). Brocadia (three species: *B. anammoxidans, B. fulgida, and B. sinica;* Oshiki et al., 2011), Kuenenia (one species: *K. propionicus*), and Anammoxoglobus (one species: *A. propionicus*).

Scalindua (three species: *S. brodae, S. sorokinii, and S. wagneri*) is a fifth genus that inhabits the marine environment in an underappreciated microdiversity (Li et al., 2011), but it is also found in freshwater and wastewater treatment systems. *Candidatus K. stuttgartiensis'* genome was reconstructed from an environmental metagenome, and genome sequencing initiatives for a number of additional species are under underway (Gori et al., 2011). The genome and transcriptome of *K. stuttgartiensis* revealed one of the most redundant organizations in terms of electron transfer pathways, with 63 distinct cytochrome c-type proteins accounting for almost 30% of their protein complement. Furthermore, genome sequencing and concurrent physiological investigations revealed anammox bacteria to be more than lithotrophic specialists, implying a new role for the bacteria.

Anammox bacteria, such as *Brocadia, Kuenenia, Anammoxoglobus, Jettenia, Anammoximicrobium moscowii, and Scalindua*, have a flexible metabolism that involves the use of a variety of electron donors (e.g., NH^{4+} and propionate) and acceptors (e.g., NO^{2-}, NO^{3-}, and $SO_4{}^{2-}$) (Rios Del Toro et al., 2018). The first five categories are widespread in wastewater treatment and freshwater systems, whereas the last is common in saline conditions like seawater and sediments (Lawson et al., 2017). Physiology and environmental conditions have a considerable impact on the spread of anammox bacteria types.

The purities of enriched anammox bacteria were 64 and 74% in the fluidized bed reactor and the sequencing batch reactor (SBR), respectively. Anammox enrichment purity of 97.6% was achieved in a membrane bioreactor (MBR) inoculated with 60–80% purity granular anammox from the first full-scale anammox reactor. Using cultured activated sludge with less than 10% anammox purity as the seed, the purity of enriched anammox bacteria could be up to 97.7%.

4.1 Cell Biology

The cell is divided into three compartments, each of which is enclosed by a bilayer of membrane. The outermost membrane encloses both the cell and the outside compartment, known as the paryphoplasm, together with a thin cell wall. There are some evidence that *K. stuttgartiensis* has an S-layer protein lattice. Although a substantial cluster of genes coding for peptidoglycan production enzymes is found in the genome of this organism, it is uncertain whether the genes are functionally expressed. Furthermore, it is unknown whether the outer membrane is chemiosmotically closed, forming the cytoplasm membrane, or is gated by porin proteins, as is the case with Gram-negative bacteria's periplasmic membrane. The nucleoid and ribosomes are situated in the riboplasm, which is surrounded by the second membrane. It's assumed that this is where the transcription, translation, and domestic machinery are kept.

In the ribosomal compartment, glycogen granules may also be seen. Larger particles resembling polyhydroxyalkanoate bodies are stored in the riboplasm of *B. fulgida* and *A. propionicus*. Anammoxosomes are a third membrane-enclosed compartment in the anammox cell design (Lindsay et al., 2001a). The anammox process, which provides energy for these bacteria's development, takes place in this prokaryotic organelle (Neumann et al., 2014). In contrast to other planctomycetes, the anammoxosome membrane is curved rather than the cytoplasmic membrane. The various membrane-associated proteins participating in the anammox process are thought to benefit from the curvature of the membrane in this circumstance (Van Teeseling et al., 2014).

Anammox bacteria use a specialized collection of cytochrome c-type proteins for catabolism, including hydroxylamine oxidoreductase-like octaheme proteins (HAOs) and hydrazine synthase (HZS). Energy should be preserved through a chemiosmotic process involving a membrane-bound ATP synthase in chemolithotrophs (ATPase). The data together strongly support the anammoxosome's role as the anammox cell's bioenergetic heart. Three lipid membrane layers divide the anammox cell into compartments. Nature's wonderful invention is the membrane, which physically and chemically separates the inside cell from the outside world. Semi-permeable membranes allow for the preservation of concentration variations in (charged) substances and the creation of energy-saving gradients. The creation of an almost infinite number of, often species-specific, lipid molecules with varied chemical and physical characteristics and combinations thereof used to improve membrane fluidity and stiffness demonstrates an amazing variety on the subject. Ladderane compounds have been introduced by Anammox bacteria. Anammox membranes, like those of all other living things, are made up of glycerolipids.

The lipids are made up of a mix of ester-linked fatty acids (found in Bacteria and Eukarya) and ether-linked long-chain alcohols (found in Archaea). The presence of saturated C17-C20 fatty acids and alcohols, which are united via cis-ring junctions to form ladder-like ("ladderane") cyclobutane and cyclohexane ring structures, distinguishes anammox. Furthermore, a proteinaceous two-dimensional crystalline surface

(S-) layer was discovered in the model species "*Candidatus Kuenenia stuttgartiensis*" (referred to as *K. stuttgartiensis* throughout the text) (Van Teeseling et al., 2014).

S-layers have been discovered in practically all evolutionary branches of Gram-positive and Gram-negative Bacteria, as well as Archaea (Fagan & Fairweather, 2014); however, this was the first time they have been discovered in a cultivated planctomycete. S-layers are thought to give cells a selective advantage by performing one or more of the following functions: osmoprotection, cell shape and integrity preservation, predation protection, and exoenzyme attachment. The intrinsic self-assembly capacity of the contributing S-layer (glyco) protein monomers forms S-layers. S-layers are frequently found with covalently connected glycans, which are exceedingly variable in terms of composition and structure. The S-layer of *K. stuttgartiensis* is made up of several copies of the modified protein Kustd1514, which has an apparent molecular mass of 250 kDa and may also be found in its unmodified form (160 kDa) to a lesser extent.

Because the S-layer is the outermost cell envelope layer and hence the structure in interface with the outside environment, it is expected to have a significant role in the cells' function. The fact that *K. stuttgartiensis* cells have not lost their S-layer after lengthy cultivating under laboratory conditions, despite the fact that this is a common occurrence for other S-layer containing bacteria, supports this notion. Because anammox bacteria lack a genetic system, a knock-out mutant to evaluate the relevance of the S-layer was not viable (Van Teeseling et al., 2014).

Anammox bacteria multiply by binary fission (Niftrik & Jetten, 2012). In primitive prokaryotes, cell division is already a sophisticated event involving several interacting protein complexes that follow a specific pattern. The emergence of a division ring in the midcell paryphoplasm under the electron microscope is the first sign of anammox cell division (Niftrik & Jetten, 2012). The cell wall invaginates somewhat after that, and the cell, including the anammoxosome, begins to lengthen, doubling in size, and constriction continues until the two cells are separated. The FtsZ protein is a typical essential participant in prokaryotic cell division. It is responsible for forming the division ring and attracting the other protein components. The genes coding for FtsZ and related "divisome" proteins are not found in the *K. stuttgartiansis* genome.

4.2 The Paryphoplasm

The paryphoplasm is the anammox cell's outermost compartment, accounting for around 20% of the total cell volume and serving an unclear purpose. The identity of the paryphoplasm as either periplasmic or cytoplasmic depends on which membrane is defined as the cytoplasmic membrane. The cytoplasmic membrane has been identified as the outermost anammox membrane. As a result, the paryphoplasm of anammox bacteria is currently treated as a cytoplasmic compartment. Lindsay et al. (2001b) speculate that the paryphoplasm contains some RNA, but no ribosomes or DNA.

The cell division ring may be seen in the paryphoplasm of dividing anammox cells, despite the fact that the contents of the paryphoplasm remain unknown. The cell division ring resembles a bracket-shaped electron-dense structure. Unlike other Planctomycetes, Anammox bacteria divide through binary fission rather than budding. The classic cell division ring protein FtsZ (Margolin, 2005) is absent from the *K. stuttgartiensis* metagenome (van Niftrik et al., 2009).

Furthermore, besides the protein D-alanine–D-alanine ligase (Ddl), which creates the peptidoglycan precursor D-alanine, none of the peptidoglycan production genes is found in the *K. stuttgartiensis* metaproteome (Kartal et al., 2011b). Finally, may genetically synthesize a glycan-free polypeptide to replace the standard peptidoglycan polymer. Anammox bacteria's outermost membrane, like that of all other Planctomycetes, is known as the cytoplasmic membrane. This is predicated on the immunogold localization of RNA within the paryphoplasm, which indicates that the paryphoplasm is a component of the cell, i.e. cytoplasm.

Apart from homologs of peptidoglycan-specific glycosyltransferase (GT) class 51 (CAZy database) (Lombard et al., 2013) required to polymerize the sugar backbone, all genes required for peptidoglycan biosynthesis are found in the genome of the anammox bacteria *Kuenenia stuttgartiensis*. Despite the absence of these GTs, it was recently shown that *K. stuttgartiensis* cells lysed with lysozyme in the presence of EDTA and penicillin G, inhibits their growth. Anammox bacteria employ nitrite as an electron acceptor in order to produce dinitrogen gas. Hydrazine (N_2H_4), an extremely poisonous "rocket fuel," and nitric oxide (NO) are the two intermediates in this process. The acetyl coenzyme A (CoA) pathway is used for carbon fixation. In anabolism, the catabolic anammox process is repeated 15 times to fix 1 molecule of carbon dioxide with nitrite as the electron donor, resulting in anaerobic nitrate synthesis (de Almeida et al., 2011).

Propionate, acetate, and formate by "*Candidatus Kuenenia stuttgartiensis*," "*Candidatus Anammoxoglobus propionicus*," "*Candidatus Scalindua*" spp., and "*Candidatus Brocadia*" spp.; methylamines by "*Candidatus Brocadia fulgida*"; and ferrous iron by "*Candidatus Kuenenia stuttgar* It has been demonstrated that, in addition to nitrite, iron and manganese oxides may be utilized as electron acceptors in "*Candidatus Kuenenia stuttgartiensis*" and "*Candidatus Scalindua*" spp. Additionally, a lack of ammonium can be alleviated by "*Candidatus Kuenenia stuttgartiensis*" and "*Candidatus Scalindua*" spp. employing formate as an electron donor to dissimulate nitrate to ammonium (DNRA) (Kartal et al., 2011a).

4.3 The Anammoxosome

According to all existing data, the anammoxosome houses the anammox cell's energy metabolism and is therefore comparable to mitochondria in eukaryotic animals. Inside the anammoxosome, catabolic processes occur that result in the synthesis of dinitrogen gas from ammonium and nitrite. The existence of an electron transport chain in the anammoxosome membrane is considered to cause the formation of a proton motive force across the membrane, which may then be utilized by ATPases to generate ATP.

The anammoxosome membrane is curved in anammox bacteria, with deep tubular protrusions of the membrane into the interior of the anammoxosome. A curved membrane has two primary advantages. Proteins can attach to the curvature selectively and thereby generate a microenvironment on the membrane, causing ion channels to preferentially localize in protrusions. A curved membrane increases the membrane surface and hence the quantity of membrane accessible for utilization by membrane-bound metabolic activities, in addition to generating a microenvironment.

The anammoxosome membrane is powered by protons translocating to the anammoxosome, and a proton motive force is formed, which promotes ATP production, according to this theory. As a result, it's quite likely that the anammox bacteria actively increase the area of the anammoxosome membrane by curving it to increase their metabolic activity (i.e., rate). It's unknown how the membrane of the anammoxosome is folded. If the anammoxosome lacks an active folding mechanism, it must be hypotonic, with the osmotic pressure folding its membrane inward to achieve osmotic equilibrium. Changes in lipid composition, the effect of integral membrane proteins and cytoskeletal proteins, and microtubule motor activity are all methods to actively bend a membrane (McMahon & Gallop, 2005).

The anammoxosome functions as a catabolism site, a diffusion barrier, and an efficient energy producer. Anammox bacteria, like other prokaryotes, have compartmentalization that is connected to metabolism and serves a specific cellular function: catabolism. The theory behind this compartmentalization is that separating membrane tasks between two distinct types of membranes offers the organism more flexibility in optimizing any membrane. The anammoxosome membrane can be employed to create and sustain a proton motive force for ATP production, as well as to keep as much of the anammox process' useful and harmful intermediates away from the rest of the cell as possible.

Because the cytoplasmic membrane is required for homeostasis, such as controlling intracellular ion concentrations and transport activities, it must be moderately flexible and permeable. The cell can overcome the problem of requiring a single membrane to be both impermeable and permeable by dividing these tasks. Using an intracytoplasmic compartment for ATP synthesis via this proton motive force would result in total control of the proton motive force's physical chemistry and thus more efficient energy transduction.

4.4 The Anammoxosome as a Potential Organelle in Cell Biology

Aside from distinct catabolic enzymes and membrane potential formation, the anammoxosome may have additional intriguing characteristics. Within the anammoxosome, tubular structures have been identified, which appear to pack the organelle in a dense mass, and these tubules are occasionally grouped in orderly arrays. These tubules might potentially serve as cytoskeletal components during cell division. Furthermore, because nucleoid DNA is frequently linked to the anammoxosome membrane, chromosomal segregation during division may include the anammoxosome. If these functions are verified, the anammoxosome will be a really multifunctional organelle of a completely new kind, with implications for cell structure and function evolution in general, as well as possible clues to how eukaryote-like division processes developed.

4.5 Energy Metabolism

However, immunogold labelling experiments employing an antibody targeting a purified protein that mimics hydroxylamine oxidoreductase (HAO) from aerobic ammonium oxidizers provided the first evidence for the anammoxosome's function in energy conservation. The genome of *K. stuttgartiensis* contains 10 homologs of HAO-like proteins, with at least one of these homologs expected to play a key role in the anammox mechanism by converting the intermediate hydrazine to dinitrogen gas. Lindsay et al. (2001b) found that the HAO-like protein was only found inside the anammoxosome, resulting in significant labelling of the anammoxosome matrix.

"*Candidatus Kuenenia stuttgartiensis*" has 10 HAO-like octaheme proteins encoded in its genome, 6 of which are highly expressed in the proteome and transcriptome. A biochemical model has been presented and recently verified in which numerous cytochrome c proteins catalyze the anaerobic oxidation of ammonium. It was discovered that "*Candidatus Kuenenia stuttgartiensis*" contained a cd1 NirS nitrite reductase capable of converting nitrite to NO utilizing both entire transcriptome and proteome data. The production and functional role of NO were investigated by inhibitor studies, which indicated that NO is an essential intermediate in anammox metabolism (Kartal et al., 2011b).

The second stage in the proposed approach included mixing NO with ammonium to create the highly reactive and volatile hydrazine intermediate. Kartal et al. (2011c) established that hydrazine is certainly changed over in vivo using a series of particular stable isotope assays. Despite the fact that no enzyme complex is known to convert NO and ammonium into hydrazine, the "Candidatus Kuenenia stuttgartiensis" gene cluster was significantly translated and expressed (kuste2859 to kuste2861). On two-dimensional gel electrophoresis and liquid chromatography-tandem mass spectrometry (LC–MS/MS), this chemical appeared as three very dominating spots. The protein complex was then purified as a 240 kDa multimer to homogeneity.

The generation of $^{29}N_2$ might be determined using ^{15}N-ammonium and ^{14}NO as substrates in a linked experiment with a very active hydrazine dehydrogenase (HDH). Hydrazine synthase is the provisional name for the protein complex encoded by the gene cluster kuste2859 to kuste2861. This N_2H_4 synthase, along with the NO reductase found in conventional denitrifying bacteria, are the only two enzymes known to be capable of forming two nitrogen atom bonds. The oxidizing capacity of nitric oxide produces the vast majority, if not all, of the nitrogen gas in Earth's atmosphere. This might support the theory that NO was one of our planet's earliest profound redox sinks (Ducluzeau et al., 2009; Garvin et al., 2009).

Hydrazine is oxidized to dinitrogen gas by a series of HDHs in anammox metabolism (i.e., kustc0694, one of the HAO-like proteins). The four electrons obtained from this oxidation are transmitted to soluble cytochrome c electron carriers, ubiquinone, the cytochrome bc1 complex (complex III), soluble cytochrome c electron carriers, nitrite reductase, and hydrazine synthase. Because a buildup of hydroxylamine (and NO) inhibits the hydrazine dehydrogenase, one of the other HAO-like proteins (kustc1061) has been postulated to act as a safety valve, converting any hydroxylamine into NO, which may be directly supplied into the hydrazine synthase.

By transferring protons from the riboplasm to the anammoxosome, the anammox process creates a proton gradient. This causes an electrochemical proton gradient from the anammoxosome to the riboplasm, which is alkaline and negatively charged in comparison to the anammoxosome. The protons are drawn from inside to outside the anammoxosome by this proton motive force, which may be utilized to drive the production of ATP catalysed by membrane-bound ATPases in the anammoxosome membrane. Protons would flow passively back into the riboplasm through proton pores created by the ATPases (with the electrochemical proton gradient, downhill). The hydrophilic ATP-synthesizing domain of the anammoxosome membrane-bound ATPases would be found in the riboplasm, whereas the hydrophobic proton-translocating domain would be found in the anammoxosome membrane. The ATP that has been produced is subsequently released into the riboplasm. The hypothesized concept and function of the anammoxosome are supported by additional experimental findings. Immunogold localization confirmed the presence of ATPases on the anammoxosome membrane (Van Niftrik et al., 2010).

The finding that all, or almost all, cytochrome c proteins are found in the anammoxosome of "Candidatus Kuenenia stuttgartiensis," as revealed by cytochrome peroxidase staining, adds to the suggested notion that the anammoxosome compartment is dedicated to energy production. Only inside the anammoxosome was staining seen, and it was most prominent in a 150-nm rim along the inner of the anammoxosome membrane, as well as in membrane curvatures (Van Niftrik et al., 2010). This indicates that the anammox enzymes are linked to or related with the anammoxosome membrane, and that they dwell on the anammoxosome side of the membrane, as suggested. The exact positions of enzymes in areas where the membrane was curled add to the concept of an energy-generating organelle with a folded membrane to boost catabolic activity. Cytochromes containing heme c (such as HAO and nitrite reductase) have only been detected in the periplasmic region in bacteria so far. The lack of peroxidase staining in the paryphoplasm compartment supports the theory that it is a cytoplasmic, rather than a periplasmic, compartment.

4.6 The Anammoxosome and Growth

Slow growth combined with respiration makes for an intriguing instance. Mitchell's chemiosmotic hypothesis predicts a lower limit for the respiratory rate: if the rate of respiratory proton extrusion falls below the rate of uncoupled proton inflow (owing to membrane leakiness), energy conservation via respiration is no longer possible. By altering membrane composition and breadth, bacteria may restrict proton leakiness to a certain extent. Nonetheless, calculations based on the data in these studies reveal that the activity of anammox bacteria in some real habitats is hard to explain using classic chemiosmotic theory. These bacteria will need a method to deal with this challenge, and the presence of intracytoplasmic compartments may play a role in that approach. The high membrane surface area of nitrifiers is thought to be used to accept more respiratory proteins (the specific activity of the proteins involved might be low). More surface area on the membrane might contribute to a faster maximal growth rate (as has been argued for methanotrophs).

A high uncoupled inflow of protons would result from a big continuous external membrane surface (due to membrane leakiness). The component lamellae appear to be arranged in individual vesicles, implying that the various lamellae are unique systems. As a result, *Nitrosococcus oceanus* may change the surface area of its active membrane in response to the availability of substrate. When there is sufficient of substrate, the entire membrane surface area is active. Only one of the internal membranes would be active when substrate constraint occurs, and the bacteria would still be able to conserve energy. As a result of the arrangement of the interior membranes, these bacteria may be able to survive with a wide variety of growth rates, from very sluggish in times of need to relatively fast in times of abundance.

Anammox bacteria have a small internal membrane surface area, therefore their maximal activity is lower than that of nitrifiers. In the case of the anammox, neither the internal membrane surface area nor the internal volume is very large. As a result, it appears that the anammox compartment was neither meant to increase substrate turnover or to retain significant quantities of substrate. Nonetheless, because the compartment houses the essential catabolic enzymes, it is most likely involved in catabolism. The anammoxosome's job would be to internalize the hydrazine pool and so reduce hydrazine losses. However, comparing the time constants for diffusion and conversion inside the anammoxosome reveals that hydrazine turnover is too slow to prevent hydrazine disappearance.

4.7 The Energy for Carbon Fixation

The electron flow in anammox bacteria's catabolism is cyclic, with the final stage (hydrazine oxidation) feeding electrons to the first two phases, which create hydrazine. Carbon fixation, on the other hand, necessitates the use of electrons. The Wood-Ljungdahl route, in particular, necessitates the use of high-reducing-power electrons. These are expected to be shuttled towards the reduction of carbon dioxide in the riboplasm via quinone or NADH and are most likely produced by hydrazine oxidation. The electrons used in carbon fixation must then be supplied so that enough electrons are available for nitrite reduction and the subsequent synthesis of hydrazine from NO and ammonium. These electrons are most likely derived from the oxidation of nitrite to nitrate, which explains why anammox bacteria usually produce modest quantities of nitrate (Strous et al., 1999). Because nitrite-derived electrons have a low reducing power, they are hypothesized to be pumped to a lower redox potential, i.e. acquire a larger reducing power, by investing energy in reverse electron transport before being employed in catabolic processes.

4.8 Metabolic Versatility of Anammox Bacteria

Anammox bacteria have long been thought of as specialists who have evolved to solely breathe ammonium and nitrite with strong affinity. Recent study (Strous et al., 2006) and genome analysis of *K. stuttgartiensis* have shown that the organisms are far more adaptable. Two SBRs were infected with the same activated sludge from which previously "*Candidatus Brocadia anamoxidans*" and "*Candidatus K. stuttgartiensis*" had been enriched to evaluate the influence of organic compounds on the performance and microbiological activities of the anammox bacteria (Kartal et al., 2007).

In the presence of excess ammonium and nitrate, as well as limited levels of nitrite, the two SBRs were given propionate (0.8 mM) and acetate (1 mM), respectively. When anammox bacteria thrived, influent nitrite and ammonia concentrations might rise from 2.5 mM to 45 mM over time. At all times, the nitrite effluent content was below detection. The intake concentrations of propionate and acetate (both 1 M) were likewise raised in conjunction with ammonium and nitrite up to 15 and 30 mM, respectively.

According to competition model estimates, if heterotrophic denitrifiers consumed all of the organic acids, a coculture would emerge with anammox cells accounting for 30–40% of the total biomass. Only if anammox bacteria were able to use organic substrates with greater affinity could a bigger proportion be predicted. In the two reactors, stable populations had established after 4 months. Anammox bacteria made up over 80% of the biomass in both reactors, but with an unexpected result: the anammox bacteria represented two different species in each: "*Candidatus Anammoxoglobus propionicus*" in the propio-nate reactor and "*Candidatus Brocadia fulgida*" in the acetate reactor (Kartal et al., 2007). The oxidation of organic substrates to CO_2 was able to be coupled with the reduction of nitrate and nitrite to nitrogen gas in cells from both reactors. When compared to the rate of ammonium oxidation by the cell cultures, the specific rates were 4–6%. The activity was shown to be solely related with the anammox fractions using percoll gradient ultracentrifugation.

Other anammox cultures' cells demonstrated the same capacity to convert organic substrates without induction, but with lower specific activity. In accordance with the carbon source during enrichment, the specific activities of propionate and acetate oxidation were greatest in *A. propionicus* and *B. fulgida*, respectively. Monomethylamine and dimethylamine, in addition to organic acids, functioned as electron donors for anammox bacteria. The organic molecules might be used as electron donors in catabolism and/or as a biosynthetic carbon source in cells.

Clearly, their contribution to nitrate reduction supports the first hypothesis. Furthermore, a ^{13}C isotopic fractionation investigation of a range of anammox lipid biomarkers revealed no significant changes between CO_2-grown cells and those grown on propionate (*A. propionicus*) or acetate (*B. fulgida*) (Rattray et al., 2008). CO_2 appears to be the primary building ingredient for lipid production in the organisms. This leaves us with the following conundrum: organic acid oxidation will, at least in part, follow the oxidative route of the acetyl-CoA pathway, whereas CO_2 fixation must follow the reductive route. Only one acetyl-CoA synthase/carbon monoxide dehydrogenase gene cluster is found in the genome assembly, whereas versatile methanogens that can operate the pathway in both directions have at least two such gene clusters.

Anammox bacteria can also function as denitrifiers. Using nitrate, nitrite, nitric oxide (NO), and nitrous oxide reductase, common denitrifiers convert nitrate to nitrogen gas via nitrite, NO, and N_2O. The use of ^{15}N labelling on physically pure *K. stuttgartiensis* revealed a new mechanism (Kartal et al., 2007). First, nitrate is converted to nitrite, which is then converted to ammonia in a single step, similar to the DNRA process. Following that, nitrite and ammonia are used as substrates for the anammox process, which produces dinitrogen gas as a byproduct. The presence of a dissimilatory nitrite reductase (NrfA) capable of executing the six-electron reduction to ammonia is a common feature of bacteria, particularly enteric bacteria. A calcium-dependent and oxygen-labile nitrite-reducing activity, properties that are typical for NrfA, could be highly enriched from anammox extracts, and a candidate gene cluster is present in the genome (Kartal et al., 2007). In *K. stuttgartiensis*, the apparent Nrf activity is rate limiting, resulting in the frequent, intermediary accumulation of nitrite. No such nitrite formation is found in *A. propionicus* and *B. fulgida* cell suspensions.

The existence of the alternate pathway appears to provide the organisms with a way to produce ammonia and nitrite from the more widely accessible nitrate. Anammox bacteria can employ Fe^{3+} and manganese oxides as electron acceptors in their metabolism in addition to nitrite and nitrate (Strous et al., 2006). Fe^{2+}, in addition to the organic molecules described, can act as an electron donor in nitrate reduction. The organisms' metabolic plasticity allows them to flourish in environments where the essential substrates, ammonium or nitrite, are scarce. It may also offer distinct species with their own ecological niche.

5 Genomics of Anammox Bacteria

The genome sequence of *Kuenenia stuttgartiensis*, the first anammox bacteria, was released in 2006. (Strous et al., 2006). The metagenome was recovered from a complex microbial community cultured in an SBR, with *K. stuttgartiensis* accounting for 74% of the total microbial population. Five supercontigs (4.2 MB) could be built in the end. The other five gaps were unable to be closed, and the magnitude of these gaps is unclear. The lack of any suspicious redundancy or missing critical genes in major metabolic processes, DNA replication, transcription, translation, and protein translocation, however, validated the genome's near completeness and accurate assembly. The 64 clusters of orthologous groups of proteins (COGs) present in all presently sequenced bacterial genomes recorded in the STRING database, with the exception of leucyl-tRNA synthetase, looked to be present, and the genome was assessed to be more than 98% complete.

5.1 *Carbon Dioxide Fixation and Respiration*

Because *K. stuttgartiensis* is the only known chemolithoautotroph among planctomycetes, it is impossible to forecast which carbon fixation mechanism it will adopt, however, 13C-carbon investigations have provided some clues (Schouten et al., 2004). All other known carbon fixation routes are either lacking or incomplete in *K. stuttgartiensis*, while the genome codes for a full acetyl-CoA (Wood-Ljungdahl) pathway (Strous et al., 2006). Two molecules of CO_2 are reduced and bonded to coenzyme A (HS-CoA) to generate acetyl-CoA in the Wood-Ljundahl pathway. The acetyl-CoA pathway is consistent with the strongly depleted carbon found experimentally in ladderane lipids of anammox bacteria (Schouten et al., 2004), as well as the activity of two key enzymes (formate dehydrogenase and carbon monoxide dehydrogenase) in cell-free extracts (Strous et al., 2006). One-carbon metabolism in *K. stuttgartiensis*, however, appears to use folate rather than methanopterin, as in acetogens; consequently, the genes are distinct from those encoding methanopterin-dependent one-carbon metabolising proteins reported in similar Planctomycetes.

Starting with the gluconeogenesis/glycolysis pathway and the tricarboxylic acid cycle as intermediate processes, acetyl-CoA is the substrate for all cell components. Except for a gene that codes for ATP citrate lyase, all of the pathways' genes are located in the genome. Because this enzyme appears to be missing, it's possible that the citric acid cycle was preferred for amino acid metabolism. The oxidation of reduced quinone or NADH produces the reducing equivalents necessary for CO_2 reduction. The genome contains clusters coding for NADH:ubiquinone oxidoreductase (complex I), a sodium-translocating NADH:quinone oxidoreductase, and a formate:quinone oxidoreductase complex, all of which support electron transport.

5.2 *Versatile Lifestyle*

The genomic findings from *K. stuttgartiensis* further corroborate the recently found anammox bacteria's varied lifestyle (Kartal et al., 2007). The core carbon metabolism was shown to be entirely reversible, and organic acid and amino acid transporters were discovered. Furthermore, in the *K. stuttgartiansis* genome, respiration is extremely redundant: at least 200 genes are estimated to be directly involved in respiration (Strous et al., 2006). Just adaptable heterotrophic bacteria like *Geobacter sulfurreducens* and *Shewanella oneidensis* have a comparable amount of redundancy, whereas the aerobic ammonia oxidizer *Nitrosomonas europaea* contains only 50 such genes. The discovery of three gene clusters encoding complex III (cytochrome bc1) and four gene clusters coding for ATP synthase complexes gives solid evidence that ATP is involved in cellular respiration.

5.3 *The Unique Anammox Metabolism*

The one-to-one combining of ammonium and nitrite into dinitrogen gas, followed by the oxidation of part of the nitrite to nitrate to create reducing equivalents for carbon dioxide fixation, is known as anammox. The following genes were found in the genome of *K. stuttgartiensis* to be involved in this metabolism: a nitrate::nitrite oxidoreductase (narGH), a cd1-type nitric oxide::nitrite oxidoreductase (nirS), and nine paralogues of hydroxylamine/hydrazine oxidoreductase (HAO/HZO).

5.4 *The 16S rRNA Gene*

The 16S rRNA gene is the most often utilized phylogenetic marker for researching microbial populations. AOB is one of the bacterial groups for which the 16S rRNA gene has been successfully amplified and analyzed. Based on their 16S rRNA gene sequences, AOB may be separated into two monophyletic lineages. Nitrosomonas (including *Nitrosococcus mobilis*) and Nitrosospira (containing Nitrosolobus and Nitrosovibrio) species make up the first lineage, which belongs to the betaproteobacteria (beta-AOB). *Nitrosococcus oceani* and *Nitrosococcus halophilus* belong to the second lineage, which is associated with the gammaproteobacteria (gamma-AOB). Because of AOB's evolutionary coherence, many polymerase chain reaction (PCR) primers and fluorescent in situ hybridization (FISH) probes have been developed. The 16S rRNA gene sequence, on the other hand, has a high degree of similarity (even more than 99% within the genus Nitrosospira). It has been proposed that sequencing the intergenic spacer between the 16S and 23S rRNA genes may be used to distinguish closely related species. Because of the differences in size (varying from 400 to

700 bp) and lesser sequence similarity (43–96%), this might be a useful supplement to 16S rRNA-based techniques.

5.5 *Functional Markers*

For ecological investigations, functional markers such as genes encoding essential enzymes engaged in a certain metabolic pathway, such as ammonia oxidation, have been suggested as an alternative. Ammonium is initially oxidized by the membrane-bound enzyme ammonia monooxygenase in aerobic AOB and AOA (Könneke et al., 2005). The oxidation of hydroxylamine to nitrite by hydroxylamine oxidoreductase is the second and energy-producing stage in AOB (HAO). Because existing genomic information does not support the existence of genes homologous to hao and other critical proteins discovered in AOB, an alternate method for channelling electrons has been proposed in AOA.

For ecological investigations, functional markers such as genes encoding essential enzymes engaged in a certain metabolic pathway, such as ammonia oxidation, have been suggested as an alternative. Ammonium is initially oxidized by the membrane-bound enzyme ammonia monooxygenase in aerobic AOB and AOA (Könneke et al., 2005). The oxidation of hydroxylamine to nitrite by hydroxylamine oxidoreductase is the second and energy-producing stage in AOB (HAO). Because existing genomic information does not support the existence of genes homologous to hao and other critical proteins discovered in AOB, an alternate method for channelling electrons has been proposed in AOA.

5.6 *Ammonia Monooxygenase Genes*

The genes encoding the enzyme AMO correspond to an operon with the form amoCAB in all known AOB (Norton et al., 2008). The operon is found in several copies in the genomes of beta-AOB (Norton et al., 2008), while only one copy has been identified for gamma-AOB (Klotz et al., 2008). Despite the possibility of employing the entire amoCAB operon for molecular investigations, only a part of the gene amoA has been employed as a molecular marker to investigate AOB diversity. This area, which is very small (approximately 450 bp) and well conserved, has been suggested to give less resolution than the 16S rRNAgene.

Since the first amoA sequence of *Nitrosomonas europaea* was published, the number of partial and full-length sequences accessible in public databases has skyrocketed. Several PCR primers have been published to amplify amoA, and a recent comparison of these primers revealed significant variances in their performance and specificity (Junier et al., 2008). The study of AMO-encoding genes was expanded to include amoC and amoB. A nested PCR method for amplifying ambient amoCAB sequences was recently developed, increasing the sensitivity of recognizing amo genes in samples with low AOB abundances. The genome projects of N. europaea, *Nitrosomonas eutropha* (Stein et al., 2007), and *Nitrosospira multiformis* (Norton et al., 2008) have shown the presence of two more conserved genes, orf4 and orf5, directly behind the amoCABoperon.

Singletons of amoC and orf4/orf5 are also encoded in the genomes of *N. europaea and N. multiformis.* It has been postulated that these unclustered copies may increase the flexibility of the ammonia catabolic inventory expression in the presence of changing ammonia concentrations (Norton et al., 2008). Two further genes, orf1 and orf5, have been found as belonging to the amoCAB operon in the gamma-AOB *N.oceani.* In beta-AOB, orf5 is homologous to orf5, however, orf1 does not have an equivalent. Both genes, amoR and amoD, are cotranscribed with amoCAB, resulting in the gamma-AOB amoRCABD operon, which is specific for gamma-AOB. So yet, none of these genes has been studied in the context of the environment.

Homologs to the amoA gene were first discovered in genomic community studies and cultured AOA, as previously indicated. Although the structure of an amoCAB operon has not been detected in AOA, homologs to amoC and amoB have been discovered (Nicol & Schleper, 2006). A number of primer sets have been created to amplify archaeal amoA, allowing AOA to be identified and quantified (Beman et al., 2008). Archaeal amoA sequences are being submitted to GenBank at a quicker rate than bacterial amoA sequences at the moment.

Ammonia-oxidizing bacteria with hydroxylamine/hydrazine oxidoreductase genes are able to survive under aerobic and anaerobic conditions. The gene encoding the octahaem cytochrome c (OCC) proteins, hydroxylamine oxidoreductase in AOB, and hydrazineoxidoreductase (HZO) in anammox, is one of the few possibly shared functional indicators for aerobic and anaerobic ammonia-oxidizing bacteria. HAO is thought to be the protein that relays electrons to the ubiquinone pool via two interacting cytochromes, c554 and cM552, and is responsible for the dehydrogenation of hydroxylamine to nitrite (the second phase of ammonia oxidation). HAO is the most well-studied functional component of aerobic ammonia oxidation due to its soluble nature.

HZO transforms hydrazine to N2 in anammox, creating the proton motive force for energy generation (Strous et al., 2006). Both HAO and HZO may oxidize different substrates (hydroxylamine and hydrazine, respectively; Hooper et al., 2005), and functional and sequence similarities have been found between the two enzymes (Klotz et al., 2008). Multiple copies of the hao gene have been discovered in beta-AOB genome sequencing efforts (Norton et al., 2008), as part of a gene cluster that also includes orf2 and cycAB (although cycB is missing in one of the copies in N. europaea and N. eutropha Norton et al., 2008). Candidatus Kueneniastuttgartiensis genome study showed eight highly divergent octahaemprotein regions as potential HZO candidates (Strous et al., 2006). Because of their highly conserved sequence across species (Klotz et al., 2008), the genes encoding the HAO/HZO proteins appear to be ideal as functional and phylogenetic biomarkers, and numerous sets of degenerate primers have recently been constructed based on the existing hao/hzo sequences.

Phylogenetic studies of the hao/HAO sequences found with a collection of cultured specimens of AOB were consistent with those previously described for the 16S rRNA gene and amoA/AmoA, indicating that the haogene might be a new molecular marker for AOB. Non-ammonia oxidizers, such as methane-oxidizing bacteria (MOB) and other microbes, also have the hao gene. Although severe conditions are necessary to avoid cross-amplification with non-AOB also carrying hao copies (e.g., MOB and the sulphur oxidizer), in silico examination of the degenerate primers used to amplify hao revealed that they are specific for AOB. pomeroyi; *Silicibacter pomeroyi*) Analysis of the hzo genes from anammox proved to be more complex than in the case of hao for AOB since in the genome of *Candidatus K. stuttgartiensis*, several possible OCC protein-encoding genes were identified (Strous et al., 2006). The phylogeny of hzo/HZO revealed the existence of three clusters of sequences. At least one of these, cluster1, is in agreement with the previously published rRNA phylogeny of anammox bacteria.

5.7 *Genes Involved in Nitrogen Metabolism*

AOB has extra genes that account for nitrogen metabolism-related enzymes. Using nonspecific primers in PCR, multiple isolates of beta-AOB were found to have the copper-containing dissimilatory nitrite reductase gene (nirK), among which were these genes. The 16S rRNA and amo A morphologies and the nirK tree topology are comparable. Although particular nirK-AOB primers have not yet been generated, our findings indicate that nirK sequences recovered from the environment may comprise sequences from ammonia-oxidizing bacteria. The gene that encodes for the big subunit of nitric oxide reductase has also been proven to be present in some AOB (norB). It is debatable if the amplification is relevant as an efficient and comprehensive marker of AOB, though, because it wasn't viable in all of the *Nitrosospira* strains investigated. Sequencing of the hzo genes from anammox for AOB proved to be more challenging than that of the hao genes because a significant number of putative OCC protein-encoding genes were identified in the genomes of *Candidatus K. stuttgartiensis* (Strous et al., 2006). The taxonomy of hzo/HZO revealed the following sequence groups. The only cluster that strongly matches the anammox bacteria's rRNA phylogeny is Cluster 1. Impressive results from the primers' tests on cultures, artificial materials, and environmental samples suggest that hzo may one day be employed as a phylogenetic and functional signature.

5.8 *Autotrophic Growth*

Although autotrophy is a typical trait of AOM and most AOB, heterotrophic growth has been seen in a number of AOB (Arp et al., 2007). The genes encoding the ribulose-biphosphate carboxylase/oxygenase (RuBisCO, cbbL, and cbbS) might be regarded molecular markers in the case of AOB. The cbb operon has been exposed to horizontal gene transfer, according to findings from AOB's genome projects (Norton et al., 2008) Genes for the 3-hydroxypropionate and citric acid cycles were discovered in AOA. The acetyl CoA carboxylase enzyme is one of the enzymes involved in CO_2 fixing in this metabolic pathway (ACCase). The gene accB, which codes for the biotin carboxylase, one of the three subunits of the ACCase, has been proposed as a possible molecular marker to study CO_2 assimilation in archaeal autotrophy. The full acetyl CoA pathway was discovered in the genome of *Candidatus K. stuttgartiensis* in anammox (Strous et al., 2006). However, no genes implicated in CO_2 absorption have yet been employed as molecular markers in anammox ecological investigations.

References

De Almeida, N. M., Maalcke, W. J., Keltjens, J. T, Jetten, M. S., & Kartal, B. (2011). Proteins and protein complexes involved in the biochemical reactions of anaerobic ammonium-oxidizing bacteria. *Biochemical Society Transactions, 39*(1), 303–308.

Arp, D. J., et al. (2002). Molecular biology and biochemistry of ammonia oxidation by *Nitrosomonas europaea*. *Archives of Microbiology, 178*, 250–255.

Arp, D. J, Chain, P. S, & Klotz, M. G (2007) The impact of genome analyses on our understanding of ammonia-oxidizing bacteria. *Annual Review Microbiology, 61*, 503–528.

Beman, J. M., Popp, B. N., & Francis, C. A. (2008). Molecular and biogeochemical evidence for ammonia oxidation by marine *Crenarchaeota* in the Gulf of California. *ISME Journal, 2*, 429–441.

Canfield, D. E., et al. (2010). The evolution and future of Earth's nitrogen cycle. *Science, 330*, 192–196.

Daims, H., et al. (2016). A new perspective on microbes formerly known as nitrite-oxidizing bacteria. *Trends in Microbiology, 24*, 699–712.

Devol, A. H. (2015). Denitrification, anammox, and N_2 production in marine sediments. *Annual Review of Marine Science, 7*, 403–423.

Ducluzeau, A.L., *et al.,* (2009). Was nitric oxide the first deep electron sink? *Trends in Biochemical Sciences, 34*, 9–15.

Fagan, R. P., & Fairweather, N. F. (2014). Biogenesis and functions of bacterial S-layers. *Nature Reviews Microbiology, 12*, 211–222.

Fuerst, J. A., & Sagulenko, E. (2011). Beyond the bacterium: Planctomycetes challenge our concepts of microbial structure and function. *Nature Reviews Microbiology, 9*, 403–413.

Fuerst, J. A., & Sagulenko, E. (2013). Nested bacterial boxes: Nuclear and other intracellular compartments in planctomycetes. *Journal of Molecular Microbiology and Biotechnology, 23*, 95–103.

Garvin, J., Buick, R., Anbar, A. D., Arnold, G. L., & Kaufman, A. J. (2009). Isotopic evidence for an aerobic nitrogen cycle in the latest Archean. *Science, 323*, 1045–1048.

Gori, F., Tringe, S. G., Kartal, B., Machiori, E., & Jetten, M. S. M. (2011). The metagenomic basis of anammox metabolism in *Candidatus 'Brocadia fulgida'*. *Biochemical Society Transactions, 39*, 1799–1804.

Junier, P., Kim, O. S., Molina, V., Limburg, P., Junier, T., Imhoff, J. F., & Witzel, K. P. (2008). Comparative in silico analysis of PCR primers suitedfor diagnostics and cloning of ammonia monooxygenase genes from ammonia-oxidizing bacteria. *FEMS Microbiology Ecology, 64*, 141–152.

Kartal, B., Kuypers, M. M., Lavik, G., Schalk, J., Op den Camp, H. J., Jetten, M. S., & Strous, M. (2007). Anammox bacteria disguised as denitrifiers: Nitrate reduction to dinitrogen gas via nitrite and ammonium. *Environmental Microbiology, 9*, 635–642.

Kartal, B., et al. (2010). Sewage treatment with anammox. *Science, 328*, 702–703.

Kartal, B., et al. (2012). Anammox–growth physiology, cell biology, and metabolism. *Advances in Microbial Physiology, 60*, 211–262.

Kartal, B., et al. (2011a). Cultivation, detection, and ecophysiology of anaerobic ammonium-oxidizing bacteria. *Methods in Enzymology, 486*, 89–108.

Kartal, B., et al. (2011b). Molecular mechanism of anaerobic ammonium oxidation. *Nature 479*, 127–130.

Kartal, B., Geerts, W., & Jetten, M. S. (2011c). Cultivation, detection, and ecophysiology of anaerobic ammonium-oxidizing bacteria. *Methods in Enzymology, 486*, 89–108

Klotz, M. G., Schmid, M. C., Strous, M., op den Cam, H. J., Jetten, M. S., & Hooper, A.B. (2008). Evolution of an octahaem cytochrome cprotein family that is key to aerobic and anaerobic ammonia oxidation by bacteria. *Environmental Microbiology, 10*, 3150–3163.

Koenneke, M., Bernhard, A. E., de la Torre, J. R., Walker, C. B., Waterbury, J. B., & Stahl, D. A. (2005). Isolation of an autotrophic ammonia-oxi-dizing marine archaeon. *Nature, 437*, 543–546.

Lam, P., Lavik, G., Jensen, M. M., et al. (2009). Revising the nitrogen cycle in the Peruvian oxygen minimum zone. *Proceedings of the National Academy of Sciences of the United States of America, 106*, 4752–4757.

Lawson, C. E., Wu, S., Bhattacharjee, A. S., Hamilton, J. J., McMahon, K. D., Goel, R., & Noguera, D. R. (2017). Metabolic network analysis reveals microbial community interactions in anammox granules. *Nature Communications, 8*, 15416.

Li, M., Cao, H., Hong, Y., & Gu, J. D. (2011). Spatial distribution and abundances of ammonia-oxidizing archaea (AOA) and ammonia-oxidizing bacteria (AOB) in mangrove sediments. *Applied Microbiology and Biotechnology, 89*, 1243–1254.

Lindsay, M. R., Webb, R. I., Strous, M., Jetten, M. S. M., Butler, M. K., Forde, R. J., et al. (2001a). Cell compartmentalization in planctomycetes: novel types of structural organization for the bacterial cell. *Archives of Microbiology, 175*, 413–429.

Lindsay, M. R., Webb, R. I., Strous, M., Jetten, M. S. M., Butler, M. K., Forde, R. J., & Fuerst, J. A. (2001b). Cell compartmentalisation in planctomycetes: novel types of structural organisation for the bacterial cell. *Archives of Microbiology, 175*, 413–429.

Lombard, V., Golaconda Ramula, H., Drula, E., Coutinho, P. M., & Henrissat, B. (2013). The carbohydrate-active enzymes database (CAZy). *Nucleic Acids Research, 42*, D490–D495.

Lu, H., et al. (2014). Microbial ecology of denitrification in biological wastewater treatment. *Water Research, 64*, 237–254.

Margolin, W. (2005). FtsZ and the division of prokaryotic cells and organelles. *Nature Reviews Molecular Cell Biology, 6*, 862–871.

McMahon, H. T., & Gallop, J. L. (2005). Membrane curvature and mechanisms of dynamic cell membrane remodelling. *Nature, 438*, 590–596.

Monteiro, M., et al. (2014). The history of aerobic ammonia oxidizers: From the first discoveries to today. *Journal of Microbiology, 52*, 537–547.

Neumann, S., Wessels, H. J. C. T., Rijpstra, W. I. C., Sinninghe Damsté, J. S., Kartal, B., & Jetten, M. S. M., et al. (2014). Isolation and characterization of a prokaryotic cell organelle from the anammox bacterium *Kuenenia stuttgartiensis. Molecular Microbiology, 94*, 794–802.

Nicol, G. W., & Schleper, C. (2006). Ammonia-oxidising *Crenarchaeota*: Important players in the nitrogen cycle? *Trends in Microbiology, 14*, 207–212.

Van Niftrik, L., van Helden, M., Kirchen, S., van Donselaar, E. G., Harhangi, H. R., Webb, R. I., Fuerst, J. A., Op den Camp, H. J., Jetten, M. S., Strous, M. (2010). Intracellular localization of membrane-bound ATPases in the compartmentalized anammox bacterium '*Candidatus Kuenenia stuttgartiensis*'. *Molecular Microbiology,* 77(3), 701–715.

Norton, J. M., Klotz, M. G., Stein, L. Y., et al. (2008). Complete genome sequence of *Nitrosospira multiformis*, an ammonia-oxidizing bacterium from the soil environment. *Applied and Environment Microbiology, 74*, 3559–3572.
Oshiki, M., Shimokawa, M., Fujii, N., Satoh, H., & Okabe, S. (2011). Physiological characteristics of the anaerobic ammonium-oxidizing bacterium '*Candidatus Brocadia sinica*'. *Microbiology, 157*, 1706–1713.
Oshiki, M., et al. (2015). Ecology and physiology of anaerobic ammonium oxidizing (anammox) bacteria. *Environmental Microbiology, 18*, 2784–2796.
Rattray, J. E., et al. (2008). Ladderane lipid distribution in four genera of anammox bacteria. *Archives of Microbiology, 190*, 51–66.
Rios-Del Toro, E. E., Valenzuela, E. I., Ramírez, J. E., López-Lozano, N. E., & Cervantes, F. J. (2018). Anaerobic ammonium oxidation linked to microbial reduction of natural organic matter in marine sediments. *Environmental Science and Technology Letters, 5*, 571–577.
Rockström, J., et al. (2009). A safe operating space for humanity. *Nature, 461*, 472–475.
Schouten, S., Strous, M., Kuypers, M. M. M., Rijpstra, W. I. C., Baas, M., Schubert, C. J., Jetten, M. S. M., & Sinninghe Damsté, J. S. (2004). Stable carbon isotopic fractionations associated with inorganic carbon fixation by anaerobic ammonium oxidizing bacteria. *Applied and Environmental Microbiology, 40*, 3785–3788.
Song, B., & Tobias, C. R. (2011). Molecular and stable isotope methods to detect and measure anaerobic ammonium oxidation (anammox) in aquatic ecosystems. *Methods in Enzymology, 496*, 63–89.
Stein, L. Y., Arp, D. J., Berube, P. M., et al. (2007). Whole-genome analysis of the ammonia-oxidizing bacterium, *Nitrosomonas eutropha* C91: Impli-cations for niche adaptation. *Environmental Microbiology, 9*, 2993–3007.
Stramma, L., Prince, E. D., Schmidtko, S., Luo, J., Hoolihan, J. P., Visbeck, M., Wallace, D. W. R., Brandt, P., & Körtzinger, A. (2012). Expansion of oxygen minimum zones may reduce available habitat for tropical pelagic fishes. *Nature Climate, 2*, 33–37.
Strous, M., et al. (1999). Missing lithotroph identified as new planctomycete. *Nature, 400*, 446–449.
Strous, M., Pelletier, E., Mangenot, S., et al. (2006). Deciphering the evolution and metabolism of an anammox bacterium from a community genome. *Nature, 440*, 790–794.
Van Teeseling, M. C. F., de Almeida, N. M., Klingl, A., Speth, D. R., den Camp, H. J. M., Rachel, R., et al. (2014). A new addition to the cell plan of anammox bacteria: "*Candidatus* Kuenenia stuttgartiensis" has a protein surface layer as the outermost layer of the cell. *Journal of Bacteriology, 196*, 80–89.
Van Niftrik, L., & Jetten, M. S. M. (2012). Anaerobic ammonium-oxidizing bacteria: Unique microorganisms with exceptional properties. *Microbiology and Molecular Biology Reviews, 76*, 585–596.
Van Niftrik, L., et al. (2009). Cell division ring, a new cell division protein and vertical inheritance of a bacterial organelle in anammox planctomycetes. *Molecular Microbiology, 73*, 1009–1019.
Vitousek, P. M., et al. (2009). Nutrient imbalances in agricultural development. *Science, 324*, 1519–1520.

Cultivation, Growth Physiology, and Chemotaxonomy of Nitrite-Oxidizing Bacteria

Gülsüm Atasoy, Esra Şentürk, and Pınar Şanlıbaba

Abstract Nitrite-oxidizing bacteria (NOB) are a small group of Gram-negative, obligate aerobes and are primarily organo and/or chemoautotrophs. They oxidize nitrite to nitrate, which is the second step of the biogeochemical nitrogen (N) cycle. Because of this characteristic, it helps in removing toxic nitrite, which is harmful to living organisms. Recent studies on obligate nitrifiers have made valuable discoveries demonstrating alternative energy metabolisms, such as the oxidation of hydrogen, sulfur, formate, and other organic compounds. NOB have different lifestyles, such as heterotrophic, lithoautotrophic, and mixotrophic, and easily adapt to large and variable environments. Besides the richness of these bacteria in moderate habitats, NOB have also been detected in extreme ecosystems, such as geothermal springs, alkaline soils, and sediments. The growth of NOB has also been observed in anaerobic environments (in the absence of nitrates or oxygen), such as deep in sludge in wastewater reservoirs or in the presence of significant concentrations of sulfur. Recognizing the natural diversity of NOB is important to understand its effects on N-cycling in natural and engineered ecosystems and to reduce greenhouse gas emissions from sewage treatment plants. Despite their huge ecological importance, we have limited knowledge of the microbiology and ecology of NOB. Their activities as the main contributors and scavengers of nitric oxides in the biosphere and their key roles in the nitrification cycle have made researchers focus again on NOB; however, developing and sustaining the growth of NOB in vitro are challenging. This section summarizes the cultivation, growth physiology, and chemotaxonomy of NOB.

Keywords Cultivation · Growth physiology · Chemotaxonomy · Nitrite-oxidizing bacteria

---

G. Atasoy · E. Şentürk · P. Şanlıbaba (✉)
Department of Food Engineering, Engineering Faculty, Ankara University, 50th Year Settlement, 06830 Gölbaşı, Ankara, Turkey
e-mail: sanlibab@ankara.edu.tr

M. P. Shah (ed.), *Anammox Technology in Industrial Wastewater Treatment*,
https://doi.org/10.1007/978-981-99-3459-1_5

1 Introduction

The most important nitrogen (N) removal strategy used in wastewater treatment plants consists of nitrification and denitrification processes (Ruiz-Martínez et al., 2018). "Nitrification", which is a two-step process, is an important step in the N cycle; in the first step, ammonia (NH_4^+) is oxidized to nitrite (NO_2^-), and in the second step, nitrite is oxidized to nitrate (NO_3^-) (Daims et al., 2016; Starkenburg et al., 2008).

$$NH_4^+ + 2e^- + O_2 + 2H^+ \rightarrow NH_2OH + H_2O,$$

$$NO_2^- + H_2O \rightarrow NO_3^- + 2H^+ + 2e^-$$

Nitrification connects the anaerobic and aerobic pathways of the N cycle and provides the electron acceptor nitrate or nitrite for the processes of anaerobic ammonium oxidation and airway ammonification (Koch et al., 2015; Park et al., 2020). Denitrification is the reduction of soil nitrate to N gases by heterotrophic bacteria (Ruiz-Martínez et al., 2018). The nitrification process has a major impact on the N cycle in soil, wastewater treatment plants, and natural ecosystems (Nowka et al., 2015; Liu et al., 2018, Shah, 2020, 2021), resulting in the removal of N from these systems (Kitzinger et al., 2018). Apart from this, nitrification also alters the soil nitrite concentration, thereby initiating niche differentiation (Nowka et al., 2015).

Nitrification, the key aerobic process of the biogeochemical N cycle, is catalyzed by ammonia-oxidizing microorganisms (AOM) (bacteria [AOB] and archaea [AOA]) and nitrite-oxidizing bacteria (NOB) (Kock et al., 2015). First, ammonia is oxidized by AOM to the intermediate substrate hydroxylamine, which is then oxidized to nitrite by hydroxylamine dehydrogenase (encoded by the *haoAB* genes); in the second step, nitrite is oxidized to nitrate by the phylogenetically diverse chemolithoautotrophic NOB via the enzyme nitrite oxidoreductase (encoded by *nxrAB* genes) (Kock et al., 2015; Cai et al., 2018; Kitzinger et al., 2018; Park et al., 2020)

$$NH_4^+ + 2O_2 \rightarrow NO_3^- + H_2O + 2H^+;\ \Delta G^{\circ\prime} = -348.9\,\text{kJ mol}^{-1}$$

Thermodynamically, the oxidation of ammonia to nitrite yields -275 kJ mol^{-1} of energy $\left(\Delta G^{\circ'}\right)$, whereas the oxidation of nitrite to nitrate yields -74 kJ mol^{-1} of energy (Daims et al., 2015). Since the theoretical numerical ratio of AOB to NOB in a nitrification system is 2:1 with respect to thermodynamics and electron transfer, AOB is the dominant organism in this process; however, the synergistic relationship between them benefits both groups (Cai et al., 2018; Yao & Peng, 2017). The growth of AOB rapidly increases in the presence of NOB, which helps prevent nitrite accumulation, while the oxidation of ammonia by AOB benefits NOB by maintaining ammonia concentrations below the toxicity threshold (Cai et al., 2018). The growth balance between NOB and AOB plays a significant role in system optimization;

therefore, the faster growth of AOB and lower nitrite oxidation rate would lead to the accumulation of nitrite as an intermediate product (Yao & Peng, 2017).

Nitrite is an important source of N for many plants and microorganisms, the electron acceptor for anaerobic respiration, and makes up about 90% of the fixed N (corresponding to 5.8 × 105 Tg) in the ocean (Koch et al., 2015; Lücker et al., 2013; Park et al., 2020). It is also an electron donor for the reduction of NAD^+ via reverse electron flow and production of ATP through oxidative phosphorylation (Starkenburg et al., 2008). However, due to the toxicity of nitrites, it is a potential threat to human health and aquatic ecosystems [concentration of >1 mg L^{-1} is toxic to aquatic organisms (He et al., 2021)], thereby requiring rapid conversion to nitrate, which is assimilated by microorganisms or plants (Bock & Wagner, 2006; Yao & Peng, 2017). Moreover, precisely sensing the interaction between NOB and AOB is crucial for optimizing nitrification in biological nutrient removal plants, as under anoxic conditions, nitrite can be converted to nitrous oxide, an ozone-depleting substance, by *Nitrosomonas* of the AOB group (Yao & Peng, 2017). Therefore, NOB has a regulatory role in the N cycle (Daims et al., 2016).

Aerobic nitrite oxidation is the second microbially-mediated part of nitrification catalyzed by NOB, which is an autotrophic, usually Gram-positive, slow-growing bacteria, considered to be the basic players of the global N and carbon (C) biogeochemical cycles (Daims et al., 2016; Nowka et al., 2015; Park et al., 2020). This group of bacteria has various ecological functions within the N cycle as they can switch between nitrite oxidation and alternative energy metabolisms, such as oxidation of sulfur, hydrogen, formate, and other organic compounds (Daims et al., 2016; Kitzinger et al., 2018). Also, they are involved in complex symbiosis with heterotrophic microorganisms and AOM apart from exploiting AOM to use urea or cyanate as an energy source (Daims et al., 2016). Under nitrification conditions, the growth of NOB is directly related to nitrite oxidation kinetics and nitrite production rate (Nowka et al., 2015). In addition, nitrification rates are closely connected to environmental parameters, such as temperature, bicarbonate, or ammonia concentration (Cai et al., 2018). Since NOB are abundant and ubiquitous prokaryotes having a critical role in reducing the detrimental effects of nitrogen pollution in the ecosystem (Boddicker & Mosier, 2018; Park et al., 2020), understanding its metabolism, limits, and diversity is crucial for managing and controlling high N in degraded ecosystems (Boddicker & Mosier, 2018).

As nitrite, a substrate for nitrite oxidation, is the end product of ammonium oxidation, nitrite oxidation generally lags behind ammonium oxidation (Ma et al., 2015). Moreover, oxidation by AOM is the first and rate-limiting step of nitrification and has been widely researched, but the same is not the case for NOB (Daims et al., 2016; Liu et al., 2018). The main reason is that NOB is perceived as physiologically restricted organisms, and its cultivation in the laboratory is more challenging and time-consuming than AOM; also, research on ammonia oxidizers has gained momentum after the discovery of AOA. Thus, nitrification research has focused only on AOB for many years (Daims et al., 2016). Because of this, knowing the genomic structure and diversity of nitrite oxidizers is significant to reconstructing the evolution of nitrite oxidation (Sorokin et al., 2012). This book chapter aims to identify and understand the breeding, growth physiology, and chemotaxonomy of NOB—the mysterious nitrifying prokaryotes.

2 Mysterious Prokaryotes: Cultivation and Growth Physiology of Nitrite-Oxidizing Bacteria

Sergei Winogradsky isolated the first pure cultures of chemolithoautotrophic bacteria, AOB and NOB, involved in the process of nitrification (Daims et al., 2016). Since the isolation of the first NOB from the genus *Nitrobacter* in 1892, the number of newly isolated NOBs has been very limited due to the difficulties in cultivating this group of bacteria under laboratory conditions and their perception as obligate chemolithoautotrophs with a limited physiological repertoire (Daims et al., 2016; Sorokin et al., 2012). In the subsequent century, although a great improvement was made by elucidating the biochemistry of the genus *Nitrobacter* (NOB) based on progressive nitrification studies, the genomic and physiological characteristics of this group of bacteria are limited to a few representatives only (Daims et al., 2016; Sorokin et al., 2012). By recognizing the natural diversity of these organisms, it may be possible to evaluate the consequences of increased human N discharge (as sewage and fertilizers) on the N cycle as well as reducing the release of the greenhouse gas nitrogen protoxide from sewage treatment plants (Sorokin et al., 2012). As the activity of NOB tends to be unstable, especially in industrial systems, the leaching of nitrite from wastewater treatment plants into natural water sources can cause major ecological damage (Daims et al., 2016). Nitrite occupies a central position in the N cycle by linking aerobic and anaerobic pathways (Alawi et al., 2007); however, if the rate of nitrite oxidation in the ecosystem decreases or ceases, it can lead to the accumulation of nitrite, resulting in adverse effects, such as hypoxia, toxicity, and loss of biodiversity (Alawi et al., 2007; Lantz et al., 2021).

The fundamental problem during the enrichment and isolation of nitrite oxidants from environmental samples is attributed to the time-consuming procedures and the inability of most NOB cultures to survive standard preservation procedures, such as freeze-drying (Off et al., 2010; Spieck & Lipski, 2011). Therefore, it is very difficult to deposit newly identified species of NOB in one of the official culture collections today. Nonetheless, as a result of a better adaptation of growth parameters to natural conditions, unprecedented strains and genera have emerged (Spieck & Lipski, 2011). Because NOB has a variety of lifestyles, such as lithoautotrophic, heterotrophic, and mixotrophic, and owing to its proliferation in marine/acidic/terrestrial ecosystems, changes in the environment composition are essential for their in vitro growth (Lücker et al., 2013; Spieck & Lipski, 2011). The growth of lithotrophic nitrite oxidizers is very slow, and the production time varies from 8 h to several days. Studies have indicated that pH, substrate, oxygen, light, and temperature affect the growth rates directly (Spieck & Bock, 2005). Since temperature is the main factor shaping the community structure of NOB, it must be maintained between –5 °C and 60 °C (Alawi et al., 2007; Spieck et al., 2020). For example, high growth temperatures of up to 60 °C are required for the cultivation of NOB isolated from hot springs (Spieck & Lipski, 2011).

The growth of the majority of NOB is maximum at an optimum pH of 7.5–8.0, the temperature of 25–30 °C, and nitrite concentrations of 30–2 mM (Spieck & Bock, 2005). Among these parameters, pH is a key consideration in niche differentiation among nitrite oxidizers. Besides the case of *Nitrobacter alkalicus*, *Nitrospira* and *Nitrobacter* species also have an optimum pH range of 7.6–8.0. In physiological experiments, a neutral pH range (7.6–7.4) is generally used for NOB cultivation (Hüpeden et al., 2016). The accumulation of cells from *Nitrotoga*, *Nitrospina,* and *Nitrospira* requires prolonged feeding procedures. Studies have indicated that higher cell amounts of *Nitrobacter* and *Nitrococcus* occur with the consumption of high nitrite concentrations (Spieck & Lipski, 2011).

Regarding their physiology, a concentration of 1 mM nitrite is required for the cultivation of the marine genera *Nitrococcus, Nitrospina,* and *Nitrospira,* whereas most species in the *Nitrobacter* genus are isolated at a higher nitrite concentration of 29 mM. In addition, the use of a 2.9 mM nitrite concentration resulted in the enrichment of *Nitrospira* in activated sludge, whereas a further tenfold dilution (0.29 mM nitrite administration) gave rise to the isolation of the new "*Candidate*" genus *Candidatus Nitrotoga* from permafrost-affected soils. Moreover, the *Nitrococcus* and *Nitrobacter* cultures achieve turbidity, which is visible to the naked eye; however, it is not the case for *Nitrospira* and *Nitrotoga* cultures as they have a high tendency for aggregation and growth is accompanied by the formation of microcolonies (Spieck & Lipski, 2011). The *nxr*B and *nxr*A genes are functional and phylogenetic markers used to detect and identify uncultured nitrite oxidizers (Daims et al., 2016). In previous studies, PCR primers targeting both the genes of the *Nitrobacter* genus were developed and tested; results indicated that *nxr*B and *nxr*A genes are useful for distinguishing closely related *Nitrobacter* strains in soil samples and pure cultures (Pester et al., 2014).

NOBs have been detected primarily in soil, cave biofilm, freshwater sources, and activated sludge (Nowka et al., 2015). The habitats of this prokaryote group include geothermal resources, soils (swamps and agricultural areas), sediments, biofilms, freshwaters (rivers and lakes), underground places (caves), engineering systems (biofilters, iron pipes, wastewater treatment plants, and drinking water systems), sea (sediments, bulk water, saltwater, corals, sponges, soda lakes, and estuaries), and hypersaline lakes (Daims et al., 2016). Even though they do not form endospores, this group of organisms is resistant to periods of reductive-energy starvation and dryness for several years; the formation-accumulation of intracellular compatible solutes facilitates the adaptation to such extreme conditions (Spieck & Bock, 2005; Spieck & Lipski, 2011). However, it is imperative that cultures be "activated", as it may take a long time for growth to resume and their metabolism to be reactivated (Spieck & Lipski, 2011).

3 Chemotaxonomy and Biochemistry of Nob

NOB are a phylogenetically diverse functional group. Based on ultrastructural criteria or cell shape, they have been classified into seven bacterial genera, *Nitrococcus, Nitrolancea, Nitrospira, Nitrospina, Nitrobacter, Nitrotoga*, and *Candidatus Nitromaritima*, which are affiliated to four bacterial phyla, *Nitrospirae, Proteobacteria* (*α-proteobacteria, γ-proteobacteria,* and *δ-proteobacteria*), *Chloroflexi,* and *Nitrospinae* (Alawi et al., 2007; Daims et al., 2016; Lücker et al., 2013). Of these genera, *Candidatus Nitromaritima, Nitrolancea,* and *Nitrotoga* have been discovered in the last decade (Boddicker, 2017). Although these prokaryotic bacteria are both slow and difficult growing organisms due to their poor energy balance, their isolation and cultivation are indispensable for assigning functions to DNA sequences obtained by molecular methods (Alawi et al., 2007; Spieck & Lipski, 2011). Among the above-mentioned bacterial groups, all are Gram-positive, except for the thermophilic *Nitrolancea hollandica* (Daims et al., 2016; Spieck et al., 2020).

The examination of bacterial shape revealed that NOB can be spiral, coke, and rod-shaped. *Nitrobacter* is a pleomorphic short rod with a polar intracytoplasmic membrane, while the genera *Nitrospina* and *Nitrospira* are long rods and spirals, respectively; unlike *Nitrobacter, Nitrospina,* and *Nitrospira* are characterized by the absence of an intracytoplasmic membrane (Spieck & Bock, 2005). On the other hand, the genus *Nitrococcus* exists in the form of coccoid cells, with tubular intracytoplasmic membranes. Some strains exhibit motility through a subpolar flagellum or a single polar flagellum (Spieck & Bock, 2005). In the bacteria belonging to the genera *Candidatus Nitromaritima, Nitrospira, Nitrotoga,* and *Nitrospina*, the key enzyme of nitrite oxidation is named "nitrite-oxidizing system" (located in the periplasmic space); however, in *Nitrobacter* (the prime NOB in freshwater systems), *Nitrococcus,* and *Nitrolancea* the key enzyme is called "nitrite oxidoreductase" (NXR) (found in intracytoplasmic and cytoplasmic membranes) (Daims et al., 2016; Lantz et al., 2021; Mundinger et al., 2019; Spieck & Bock, 2005). Based on the phylogeny and cellular localization of these NXR enzymes, they serve as specific functional markers for NOB and ensure oxidation of nitrite to nitrate (Mundinger et al., 2019; Pester et al., 2014).

Studies have demonstrated the independent evolution of the periplasmic and cytoplasmic enzyme types, leading to notable phylogenetic variation among nitrite oxidizers (Daims et al., 2016; Pester et al., 2014). Cytoplasmic NXR enzymes are phylogenetically linked to membrane-bound cytoplasmic nitrate reductases, while periplasmic NXR enzymes are known to be the members of the type II dimethyl sulfoxide reductase family (Daims et al., 2016; Park et al., 2020). The cytoplasmic NXR enzyme consists of three subunits: catalytic alpha (NxrA), electron-bearing beta (NxrB), and gamma (NxrC) (Daims et al., 2016; Mundinger et al., 2019; Pester et al., 2014). The α-subunit has a substrate-binding site, whereas the β- and γ-subunits play an important role in electron transfer in the respiratory chain; apart from this, the γ-subunit acts as membrane anchor of the holoenzyme (Mundinger et al., 2019; Pester et al., 2014).

The activity of NXR is defined according to Michaelis–Menten kinetics based on V_{max} (maximum nitrite oxidation activity) and K_m (semi-saturation constant). In 1999, based on these parameters, Schramm et al. categorized *Nitrospira* to be *K*-strategist (initial selection for competitive ability in large populations) and *Nitrobacter* to be *r*-strategist (selection for high population growth in non-populated populations). At high substrate concentrations, *r*-strategists are known to exhibit maximum specific growth rate (to nitrite oxidation activity) and low substrate utilization rates, whereas the reverse is observed with *K*-strategists (Jacob et al., 2017; Nowka et al., 2015). The validity of these two strategies has not been evaluated with pure cultures; however, it has been evaluated for reactor systems via biotechnological approaches (Nowka et al., 2015). The K_m values of *Nitrobacter* strains range from 49 to 544 μM NO_2^-; conversely, for *Nitrospira* strains, it varies between 9 and 27 μM NO_2^- (Daims et al., 2016). *Nitrococcus* has the highest K_m with 120 μM NO_2^-, whereas *Nitrospina* has a K_m value of 19 μM NO_2^- (Jacob et al., 2017). For various species of *Nitrotoga* genus, V_{max} and K_m values have been reported to be in the range of 52–19 μM NO_2^-/hr mg protein and 25–89 μM NO_2^-, respectively (Spieck et al., 2021). *"Cand. Nitrotoga arctica"* and *Nitrolancea hollandica* have been categorized as *r*-strategists, with a K_m constant of 58 μM and 1 mM, respectively (Daims et al., 2016).

3.1 Proteobacteria Bacterial Phylum

3.1.1 Genus of Nitrobacter

Nitrobacter genus, the primary model organism for studying nitrite oxidation, was first isolated by Winogradsky in 1982; however, successive studies have pointed toward the limited number of newly isolated nitrite oxidizers due to the difficulty in cultivating members of this genus under laboratory conditions (Park et al., 2020; Starkenburg et al., 2006). *Nitrobacter* spp., which oxidizes nitrite to nitrate with the help of the NXR enzyme, fixes CO_2 by means of the Calvin-Benson-Basham pathway (Spieck & Bock, 2005; Starkenburg et al., 2006, 2008). It also has the ability to directly assimilate organic C compounds, such as acetate, pyruvate, formate, glycerol, and α-ketoglutarate, in the absence of nitrite (Starkenburg et al., 2008), thereby exhibiting heteroorganotrophic growth ability. However, most of the other NOBs cannot assimilate C from organic compounds (Spieck & Lipski, 2011). Nonetheless, in some strains of this genus, it has been observed that the combination of organic C sources and nitrite can outpace chemolitho- or chemoorganotrophic growth (Starkenburg et al., 2006).

Nitrobacter, a member of *Alphaproteobacteria,* is divided into four species: *Nitrobacter alkalicus* (predominantly found in soda lakes), *Nitrobacter winogradskyi* (mainly available in marine, brackish waters, and coastal shrimp recirculating aquaculture system), *Nitrobacter vulgaris* (mainly available in freshwater aquarium biofilter, brackish and freshwater, and marine water), and *Nitrobacter hamburgensis* (dominant in freshwater sediments) (He et al., 2021). *Nitrobacter winogradskyi* strain obtains the highest growth rates using nitrite as the energy source, but *Nitrobacter hamburgensis* X14 strain achieves the highest growth rates in the presence of nitrite and organic C. Therefore, under nitrite depletion conditions, organic C additions to nitrite-containing cultures can increase protein content and growth efficiency of NOB (Starkenburg et al., 2008).

Nitrobacter are primary prokaryotes generally found in terrestrial and freshwater habitats; they are not dominant in ecosystems with low ambient nitrite levels (unfertilized soil, wastewater treatment plant, etc.) as their growth requires high nitrite concentrations (Koch et al., 2015; Lantz et al., 2021; Park et al., 2020; Spieck & Bock, 2005). Because of the accumulation of extracellular compatible solutes, they are resistant to both reductive-energy starvation and periods of dryness (Spieck & Bock, 2005; Spieck & Lipski, 2011). For example, *Nitrobacter vulgaris*, the dominant species in various habitats, can survive for 24 months without water. In addition, studies have demonstrated that this bacterial species can accumulate sucrose, betaine, and glycine amino acids in the environment and also produce trehalose sugar. The Gram-negative, facultative chemolithoautotroph *Nitrobacter* can also tolerate changing environmental conditions (Spieck & Bock, 2005; Starkenburg et al., 2008) as members of this genus can also be found in sewage and marine environments. Most of the known purified *Nitrobacter* strains have been obtained from natural stones, freshwater rivers, desert soils, sand filters of waterworks, and unusual environments, such as sulfidic ore mines (He et al., 2021; Spieck & Bock, 2005). *Nitrobacter winogradskyi* species have been detected in habitats like freshwater, soil, concrete sewage, whereas *Nitrobacter hamburgensis* species, which has the largest *Nitrobacter* genomes, are found only in soil (Spieck & Bock, 2005; Starkenburg et al., 2011). So far, only *Nitrobacter winogradskyi* ZS-1 strain has been isolated from brackish aquaculture water, whereas the isolation of *Nitrobacter* strains from freshwater aquaculture waters has not been reported yet (He et al., 2021). *Nitrobacter winogradskyi* is often used as a model NOB due to its superior growth efficiency and nitrite tolerance compared to other NOBs (Mellbye et al., 2015). *Nitrobacter winogradskyi* strain has been cultured aerobically in 60–29 mmol L^{-1} $NaNO_2$ medium at 28–30 °C (Zhang et al., 2018). Sequencing of the *Nitrobacter winogradskyi* genome revealed a single circular chromosome of 3,402,093 bp encoding 3,143 predicted proteins (Starkenburg et al., 2006).

In terms of 16S rRNA gene sequencing (98% to 97% identity of 16S rRNA), *Nitrobacter* is a widely distributed NOB group within the *Bradyrhizobacaea* family, and strains of this genus are closely related to *Bradyrhizobium japonicum* and the metabolically versatile phototrophic bacterium *Rhodopseudomonas palustris* (Mellbye et al., 2015; Starkenburg et al., 2006, 2008). Among the close relatives, *Bradyrhizobium japonicum* has one of the largest prokaryotic genomes. *Rhodopseudomonas. palustris* in the *α-proteobacteria* can also develop chemotrophically. However, none of these close relatives use nitrite as an energy source (Starkenburg et al., 2008).

3.1.2 Genus of Nitrococcus

The genus *Nitrococcus* is a member of the family *Ectothiorhodospiraceae*, order *Chromatiales*, clade *Gammaproteobacteria*. These prokaryotes are obligate lithoautotrophs, Gram-negative, aerobic, and obtain energy by oxidizing nitrite with the help of the NXR enzyme. They have been isolated from marine environments, exhibiting optimum growth in 70–100% seawater enriched with nitrite and other inorganic salts (Spieck and Bock, 2015). They can also utilize organic matter as C and energy source (Füssel et al., 2017). Three identified effects of NOBs on aquatic habitats are: (1) they are responsible for nitrate formation, (2) they provide the substrate for denitrification, and (3) they reduce nitrite toxicity to both fish and other aquatic organisms (Lantz et al., 2021).

Nitrococcus and a member of the *Deltaproteobacteria* genus *Nitrospina* are the major nitrite oxidizers associated with marine ecosystems (Koch et al., 2015; Spieck & Bock, 2005). Till now, very few representatives of marine NOB cultures (*Nitrococcus mobilis, Nitrospina gracilis/watsonii,* and *Nitrospira marina/ecomares*) have been reported (Park et al., 2020). The only known species, *Nitrococcus mobilis*, has been reported three times so far. It was detected twice in the Pacific Ocean in the 1980s and once in a high-salinity pond. Most recently, it was reported in the Namibian OMZ (oxygen minimum zones), where they even surpassed *Nitrospina*. Its natural presence in the waters of the OMZ demonstrates the use of alternative anaerobic metabolisms (Füssel et al., 2017).

Füssel et al. (2017) have analyzed the genome of *Nitrococcus mobilis* Nb-231. According to them, the *Nitrococcus mobilis* genome can encode duplicated versions of the NXR α-subunit (NxrA). The catalytic subunit NxrA1 and the β-and γ-subunits are encoded in one genomic region. Moreover, unlike other nitrite-oxidizing prokaryotes, *Nitrococcus mobilis* can encode a copper-containing nitrite reductase (NirK). It is assumed that the NirK enzyme facilitates the reduction of nitrite to nitric oxide (NO). The *Nitrococcus mobilis* genome also influences the sulfur cycle as it contains several key sulfur-metabolizing enzymes, including a sulfide/quinone oxidoreductase that can oxidize H_2S to elemental sulfur or polysulfide in the periplasmic space (Füssel et al., 2017).

3.2 Chloroflexi Bacterial Phylum

3.2.1 Genus of Candidatus Nitrotoga

Due to recent advances in culturing and non-culturing approaches, several new lineages have been added to the phylogenetic diversity of NOB, such as the genus *Nitrolancea* in *Chloroflexi*, the candidate genus "*Candidatus Nitrotoga*" in *Betaproteobacteria,* and the candidate genus "*Candidatus Nitromaritima*" in *Nitrospinae* (Kitzinger et al., 2020; Park et al., 2020). "*Cand. Nitrotoga*" was first isolated from activated charcoal (Kitzinger et al., 2018). Only four *Nitrotoga* cultures have been reported so far, and their genome sequences are not available (Boddicker & Mosier, 2018). This genus was named "toga" owing to the irregularly shaped cocci or short rods with an oddly large periplasmic space. They live in microcolonies, without the formation of biofilms (Spieck et al., 2021).

The *Candidatus Nitrotoga* genus has contributed remarkably to nitrification research in recent years because of its presence in engineered and natural ecosystems, including wastewater treatment plants (Boddicker, 2017; Kitzinger et al., 2018). Based on recent molecular research, it has been reported to have a surprisingly wide range of habitats, including Antarctic subglacial lakes, glacial soils, underground cavern, tap water, salt marsh sediments, antibiotic-rich sediments, Yellow Sea, and rivers (Boddicker, 2017; Boddicker & Mosier, 2018; Kitzinger et al., 2020; Lantz et al., 2021). Also, it is not only limited to cold climates but also globally distributed in a wide temperature range, from the tropics to the poles (Spieck et al., 2021). In freshwater habitats, *Candidatus Nitrotoga*-like sequences constitute approximately 0.01–10% of the sequences of the total bacterial community (Lantz et al., 2021). On the other hand, it is the only NOB group observed in subglacial lakes and wastewater treatment plants, forming 2–13% of the total bacterial community (Boddicker & Mosier, 2018). In addition, optimum temperatures for the physiological activity of this genus vary between 10 and 28 °C. In spite of this, *Nitrotoga* has been detected at temperatures as low as 4–17 °C. Thus, low temperatures facilitate the growth of *Nitrotoga,* as most cultures have been reported to develop well at 4 °C (Spieck et al., 2021). Moreover, physiological studies have demonstrated that the pH range of 6.8–8.3 is apt for the cultivation of *Candidatus Nitrotoga.* Studies on *Nitrospira* genus indicate that high-nitrite conditions support the growth of *Nitrotoga* (Lantz et al., 2021; Spieck et al., 2021), as the growth of *Nitrotoga* in low-nitrite conditions is quite slow, with production times of 44–54 h. *Candidatus Nitrotoga* can be frozen by a special method (cryopreservation); however, accumulation in bacterial culture collections is not possible because the reactivation of the cultures requires intense manpower. Also, the activation is less successful than other nitrite oxidizers. Consequently, *Nitrotoga* is still known as a *Candidatus* taxon as no strains of this genus have been validated yet (Spieck et al., 2021).

In 2007, the isolation of *Candidatus Nitrotoga arctica* belonging to the *β-proteobacteriales* order from Siberian permafrost soil, *Candidatus Nitrotoga* sp. AM1 from coastal sand, and *Candidatus Nitrotoga* sp. HAM-1 from activated sludge was performed (Boddicker & Mosier, 2018; Ishii et al., 2020). *Nitrotoga arctica* has moderate nitrite affinity, as it adapts well to nitrite concentrations of 0.3 mM; also, the optimum temperature for its growth is 10–17 °C (Boddicker & Mosier, 2018; Spieck et al., 2021). The optimum pH range for this group of bacteria is 6.4–6.8, as it is abundant in slightly acidic environments (Spieck et al., 2021). However, it was determined that *Nitrotoga fabula* isolated from a wastewater treatment plant in Austria exhibited poor growth at temperatures below 20 °C; the optimum growth was observed between 24 and 28 °C, and it indicated a high tolerance to nitrate and nitrite (Kitzinger et al., 2018; Spieck et al., 2021).

Candidatus Nitrotoga-like 16S rRNA genes have been detected in glacial and periglacial soils, indicating that they are the only cold-adapted nitrite oxidizers (Ishii et al., 2020). Given the worldwide abundance and distribution of *Nitrotoga* species, these prokaryotes are predicted to play a critical role in the global N cycle. However, despite their importance in the ecosystem, there is limited data on their microbiology as no pure cultures and genomes exist (Boddicker et al., 2017; Kitzinger et al., 2018).

Nitrotoga can strangely coexist with the genera *Nitrospira* and *Nitrobacter* in a N-limiting ecosystem (Ishii et al., 2020; Spieck et al., 2021). In 20 full-fledged wastewater treatment plants in Switzerland and Germany, both *Nitrospira* and *Nitrotoga* were detected in about half of the activated sludge; *Nitrotoga* species were also identified in both the plants (Hüpeden et al., 2016; Ishii et al., 2017). Furthermore, *Nitrotoga* was even more abundant than *Nitrospira*, which is predominantly found in activated sludge in many Danish plants (Hüpeden et al., 2016). Phylogenetic analyses of NXR for nitrite oxidation have demonstrated that the oxidation capacity in *Nitrotoga* has evolved independently of other nitrite oxidizers (Ishii et al., 2020). That is, it has a different type of periplasmic NXR, and its K_m values were observed to be in the range of 25–89 μM NO_2^- (*K*-strategist). The V_{max} activity of *Nitrotoga* was observed to be in the range of 52–19 μmol nitrite/hr mg protein (Spiecek et al., 2021).

3.2.2 Genus of Nitrolancea

Nitrolancea is a group of bacteria within the family *Sphaerobacteriaceae*, the order *Sphaerobacteriales,* the class *Thermomicrobia*, the phylum *Chloroflexi,* and this nitrite-oxidizing group oxidizes formate to CO_2 as an energy source (Sorokin et al., 2018). Thus, it utilizes CO_2 as the C source and ammonium as the N source. Also, it is neutrophilic, highly nitrite tolerant, and thermotolerant (Sorokin et al., 2014). Bacteria belonging to this phylum are thermophilic, and *Nitrolancea hollandica* is the only species determined (Sorokin et al., 2018; Spieck et al., 2020). Recent studies have identified novel thermophilic nitrite oxidizers belonging to the *Chloroflexi* phylum from the hot springs of Yellowstone National Park. Besides, 16S rRNA gene sequences associated with the genus *Nitrolancea* have been detected in warm environments, such as the Kalahari and Atacama Desert (Spieck et al., 2020).

Nitrolancea hollandica, unlike other nitrite oxidizers, has a Gram-positive cell wall (contains lysine and ornithine amino acids) (Daims et al., 2016; Sorokin et al., 2014). *Nitrolancea hollandica* LbT strain has been isolated from a laboratory-scale nitrification bioreactor operated at 35 °C with a high ammonium bicarbonate loading rate (Sorokin et al., 2018; Spieck et al., 2020). It has also been suggested that *Nitrolancea*-like 16S rRNA gene sequences are rarely detected in specialized incubations of soil or water samples, and the habitat of *Nitrolancea hollandica* has high ammonia and nitrite concentrations, as well as high temperatures (Spieck et al., 2020). The maximum and optimum growth temperatures of this species are 46 and 40 °C, respectively; it develops between optimum pH of 6.8–7.5 (Sorokin et al., 2014). The substrate affinity (K_m (NO_2^-)) of *Nitrolancea hollandica* is low (1 mM); however, the optimum nitrite concentration varies between 5 and 20 mM (Sorokin et al., 2014; Spieck et al., 2020). In addition, the tolerable nitrite concentration is observed to be 75 mM (Sorokin et al., 2014). However, no data are available on the minimum temperature limit for the proliferation of this group of nitrite-oxidizing prokaryotes (Spieck et al., 2020).

3.3 *Nitrospinae Bacterial Phylum*

3.3.1 Genus of Nitrospira

Among all NOBs, the genus *Nitrospira* is reported to be more versatile both metabolically and ecophysiologically (Park et al., 2020). The metabolic versatility, such as providing an economical route for nitrite oxidation, using different organic compounds, and using ammonia released from urea or cyanate, has been influential on the ecological success of this genus (Daims et al., 2015).

This genus belongs to the *Nitrospirae* phylum, *Nitrospira* class, *Nitrospirales* order, and *Nitrospiraceae* family. *Nitrospira marina* is the first species to be isolated from water collected from the Gulf of Maine by Watson et al. (1986), Daims and Wagner (2018). Till now, only four pure cultures of *Nitrospira* genus have been isolated, namely *Nitrospira calida, Nitrospira marina, Nitrospira moscoviensis,* and "*Candidatus Nitrospira bockiana*" (retrieved from activated sludge) (Lebedeva et al., 2008; Off et al., 2010). Cultures isolated from activated sludge consist of planktonic cells with very small clusters (4–3 μm), small micro-colonies (12–1 μm), or large cell clusters in the range of 40–600 μm (Mehrani et al., 2020). Selected species of this genus are obligate lithotrophs capable of mixotrophic growth (Lebedeva et al., 2008). Studies have demonstrated that most of the *Nitrospira* strains form biofilm structures and proliferate intensively in microcolonies (Mehrani et al., 2020). It occurs in the rhizosphere, leaf, and root surface of plants and colonizes sea sponges (Daims & Wagner, 2018).

Aerobic chemolithoautotrophic *Nitrospira* has a very substantial role in the process of nitrification as it is closely related to AOA and AOB (Daims & Wagner, 2018). Moreover, members of this recently discovered genus are named "comammox" organisms or "complete ammonia oxidizers" since they alone catalyze both nitrification steps, thereby altering the current understanding of nitrogen removal processes (Daims & Wagner, 2018; Mehrani et al., 2020). Comammox organisms employ ammonia monooxygenase and hydroxylamine dehydrogenase enzymes to oxidize ammonia to nitrite via hydroxylamine intermediate; the CO_2 is fixed by the reducing citric acid cycle.

$$NH_4^+ + 2O_2 \rightarrow NO_3^- + H_2O + 2H^+;\ \Delta G^{\circ'} = -348.9J^{-1}$$

It has been reported that *Candidatus Nitrospira inopinata*, a comammox strain isolated from hot groundwater, has a very high affinity for ammonia (Daims & Wagner, 2018). The genome of *Candidatus Nitrospira inopinata* has *nxr*A and *nxr*B genes encoding the α-and β-subunits and genes encoding the *Nxr*C gamma subunit of the NXR enzyme (Daims et al., 2015).

All known comammox organisms of the genus *Nitrospira*, classified into at least six phylogenetic lineages by comparative analysis of 16S rRNA gene sequences, belong to lineage II; these organisms are widely found in engineered and natural ecosystems (Daims & Wagner, 2018; Daims et al., 2015; Ushiki et al., 2013). *Nitrospira* IV, II, and I strains have been detected in wastewater treatment plants; however, predominantly, strain I or II is observed in both full-scale treatment plants and laboratory systems (Mehrani et al., 2020). *Nitrospira lenta, Nitrospira moscoviensis,* and *Nitrospira japonica* are the other cultured *Nitrospira* strains in line II. These strains lack the enzymatic repertoire for using ammonia as an energy source; however, the opposite is the case for *Nitrospira inopinata* in strain II. The *nxr*B gene serves as a phylogenetic marker to differentiate *Nitrospira* strains (Daims et al., 2015). In addition, Lipski et al. (2001) detected three essential fatty acids that cis 11 and cis 7 hexadecenoic acid and 11-methyl-hexadecanoic acid isomers are characteristic of *Nitrospira* species. These fatty acids are chemotaxonomic markers used for the identification of *Nitrospira* species in engineered and natural ecosystems (Keuter et al., 2011).

This phylogenetically diverse genus is observed in many natural and engineered ecosystems such as aquaculture biofilters, soil and hot springs, freshwater systems, geothermal springs, wastewater treatment plants, and ocean (Daims & Wagner, 2018; Koch et al., 2015; Lebedeva et al., 2008; Off et al., 2010). It has been reported that *Nitrospira* or *Nitrobacter* species are generally responsible for the oxidation of nitrite in wastewater treatment plants; however, most of the studies are focused on *Nitrobacter* strains as they can be easily cultured in the laboratory (Ushiki et al., 2013). Because of various immunological techniques and cultivation-independent molecular methods, *Nitrospira* species, in contrast to *Nitrobacter* species, have been determined to be dominant in various municipal wastewater treatment plants or aquariums (Mehrani et al., 2020; Ushiki et al., 2013). Members of this genus are the dominant nitrite-oxidizing bacteria in wastewater treatment plants (Koch et al., 2015), and operational dissolved oxygen concentration in these plants affects *Nitrospira* abundance. Apart from dissolved oxygen and low nitrite concentrations, members of this genus can adapt to high pH (8–8.3) and temperatures (29–30 °C) (Mehrani et al., 2020). Some species in activated sludge or marine ecosystems utilize alternative organic substrates, such as glycerol, H_2, formate, and pyruvate, as alternative energy sources whenever they are limited to a single reducing and energy source (Daims & Wagner, 2018; Koch et al., 2015; Park et al., 2020). Due to its metabolic diversity, *Nitrospira* adapts to different environmental conditions and colonizes distinct habitats (Daims & Wagner, 2018). For example, *Nitrospira moscoviensis* species, isolated from heating pipes, grow under aerobic H_2 oxidation conditions (Daims & Wagner, 2018; Koch et al., 2015). The optimum temperature for most of the NOB is 28 °C, whereas the nitrite oxidizer *Nitrospira moscoviensis* grows optimally at 39 °C (Alawi et al., 2007).

Many strains of the genus *Nitrospira* have not been cultured yet, and since a few isolates are slow to grow in the laboratory, it remains difficult to grow sufficient biomass for physiological studies (Daims & Wagner, 2018; Mehrani et al., 2020). Thus, the effect of growth parameters, physiology, environmental conditions, and inhibitory compounds on *Nitrospira* activity still remains a mystery (Mehrani et al., 2020). The first sequenced genome, consisting of a circular chromosome of 4.3 Mbp, has been obtained from *Nitrospira defluvii* isolated from activated sludge. Plasmids have not been detected in any of the bacterial strains belonging to this genus (Daims & Wagner, 2018).

3.3.2 Genus of Nitrospina

Though abundant in ocean and coastal ecosystems, the habitat range and ecophysiology of the genus *Nitrospina* remain largely a mystery due to the lack of knowledge of the cultured representatives (Park et al., 2020; Yepsen et al., 2019). Therefore, the phylogenetic relationship of this genus is still unclear. The obstinacy of bacteria belonging to the genus *Nitrospina* against cultivation in the laboratory and their slow growth rate (24 h production time) further challenges the determination of biochemical characterization (Lücker et al., 2013). The genus *Nitrospina* is mainly restricted to marine environments, as its non-marine habitats include rare places, such as uranium mill residues and an alpine underground radioactive thermal source (Lücker et al., 2013). The majority of members of this genus have been detected in the surface and deep ocean, including oxic/anoxic sediments, deep estuarine environments, and OMZs (Yepsen et al., 2019), which are marine areas where oxygen cannot be detected at medium depths. Within these regions, the oxygen-depleted zone (ODZ) induces anaerobic microbial processes, leading to the constant loss of N through anammox and denitrification. In spite of being aerobic, the genus *Nitrospina* oxidizes nitrite to nitrate in ODZs. Since no anaerobic nitrite oxidizer has been identified so far, how oxidation happens in this anoxic environment remains a mystery (Sun et al., 2019). The fact that the distribution profiles of archaeal 16S rRNA and *Nitrospina*-like 16S rRNA genes have correlated in some coastal and open ocean habitats could throw some light on this mystery (Lücker et al., 2013).

Nitrospina gracilis strain 211, the first isolate of the genus *Nitrospina*, was identified in 1971 in the Atlantic Ocean surface water. Bacteria belonging to this genus lack intracytoplasmic membranes and differ from candidate genera *Candidatus Nitrotoga* and *Nitrospira* due to the absence of the characteristic wide periplasmic space (Spieck et al., 2014). To date, only *Nitrospinae* species distantly related to the environmentally abundant uncultured *Nitrospinae* phylum have been isolated (Mueller et al., 2021; Sun et al., 2019). Only two species (*Nitrospina watsonii* and *Nitrospina gracilis*) have been cultured in high O_2 concentrations (Sun et al., 2019).

Nitrospina gracilis forms Gram-negative rods (Lücker et al., 2013; Spieck et al., 2014) and grows chemoautotrophically using nitrite as the sole energy source and CO_2 as the C source. It is aerobic as it utilizes O_2 as the terminal electron acceptor (Lücker et al., 2013). This bacterium contains a full set of genes for reverse electron transport from nitrite to NADH (Spieck et al., 2014). Species belonging to this genus have been detected both in marine OMZs and anoxic marine sediments by molecular methods (Lücker et al., 2013). It has no heterotrophic potential and exploits the reductive tricarboxylic acid cycle (TCA) for C fixation and can store fixed C as glycogen inside the cell (Spieck et al., 2014). Besides *Nitrospina gracilis*, the only cultivated species of the genus *Nitrospina* is *Nitrospina watsonii*, isolated from the 100 m deep suboxic region of the Black Sea. Both of these species are phylogenetically closely related (Ngugi et al., 2016), as genomic studies have demonstrated that the 16S rRNA gene sequence of *Nitrospina watsonii* is more than 97% identical to that of *Nitrospina gracilis* (Sun et al., 2019).

4 Conclusion

Nitrification, a significant process of biological wastewater treatment and biogeochemical N cycle, is catalyzed by physiologically versatile, Gram-negative, chemolithoautotrophic, aerobic nitrite-oxidizing bacteria (NOB) (Boddicker & Mosier, 2018; Daims et al., 2016; Kitzinger et al., 2018). The second step of nitrification, i.e., the oxidation of nitrite to nitrate, serves as the energy source for NOB (Ruiz-Martínez et al., 2018). In addition, they utilize alternative energy metabolisms, such as the oxidation of organic C, sulfur, hydrogen, and other organic compounds, for their development (Boddicker & Mosier, 2018; Kitzinger et al., 2018). The nitrification process takes place both in artificial (wastewater treatment plants) and natural (terrestrial and aquatic ecosystems) environments (Nowka et al., 2015). In spite of N removal being a fundamental process in ecosystems, our knowledge of this functional guild is limited (Spieck et al., 2020). The main reasons for this situation are that NOB are difficult to cultivate and develop slowly; therefore, enriching and isolating them from environmental samples are challenging (Off et al., 2010).

References

Alawi, M., Lipski, A., Sanders, T., & Spieck, E. (2007). Cultivation of a novel cold-adapted nitrite oxidizing betaproteobacterium from the Siberian Arctic. *The ISME Journal, 1*(3), 256–264. https://doi.org/10.1038/ismej.2007.34

Bock, E., & Wagner, M. (2006). Oxidation of inorganic nitrogen compounds as an energy source. In M. Dworkin, S. Falkow, E. Rosenberg, K.-H. Schleifer, e. stackebrandt (eds.), the prokaryotes, volume 2: ecophysiology and Biochemistry. New York, pp. 457–495. https://doi.org/10.1007/0-387-30742-7_16

Boddicker, A. M. (2017). Cultivation and genomic sequencing of novel nitrite-oxidizing bacteria. Master's Thesis. University of Colorado at Denver, pp. 1–47.

Boddicker, A. M., & Mosier, A. C. (2018). Genomic profiling of four cultivated Candidatus Nitrotoga spp. predicts broad metabolic potential and environmental distribution. *The ISME Journal, 12*(12), 2864–2882. https://doi.org/10.1038/s41396-018-0240-8

Cai, M., Ng, S. K., Lim, C. K., Lu, H., Jia, Y., & Lee, P. K. (2018). Physiological and metagenomic characterizations of the synergistic relationships between ammonia-and nitrite-oxidizing

bacteria in freshwater nitrification. *Frontiers in Microbiology, 9*, 280. https://doi.org/10.3389/fmicb.2018.00280
Daims, H., Lebedeva, E. V., Pjevac, P., Han, P., Herbold, C., Albertsen, M., & Wagner, M. (2015). Complete nitrification by Nitrospira bacteria. *Nature, 528*(7583), 504–509. https://doi.org/10.1038/nature16461
Daims, H., Lücker, S., & Wagner, M. (2016). A new perspective on microbes formerly known as nitrite-oxidizing bacteria. *Trends in Microbiology, 24*(9), 699–712. https://doi.org/10.1016/j.tim.2016.05.004
Daims, H., & Wagner, M. (2018). Nitrospira. *Trends in Microbiology, 26*(5), 462–463. https://doi.org/10.1016/j.tim.2018.02.001
Füssel, J., Lücker, S., Yilmaz, P., Nowka, B., van Kessel, M. A., Bourceau, P., & Lam, P. (2017). Adaptability as the key to success for the ubiquitous marine nitrite oxidizer Nitrococcus. *Science Advances, 3*(11), e1700807. https://doi.org/10.1126/sciadv.1700807
He, M., Xiong, Y., & Cheng, K. (2021). Characters of a nitrobacter enrichment culture from a freshwater aquaculture pond. *Biotechnology and Biotechnological Equipment, 35*(1), 1343–1352. https://doi.org/10.1080/13102818.2021.1974944
Hüpeden, J., Wegen, S., Off, S., Lücker, S., Bedarf, Y., Daims, H., & Spieck, E. (2016). Relative abundance of Nitrotoga spp. in a biofilter of a cold-freshwater aquaculture plant appears to be stimulated by slightly acidic pH. *Applied and Environmental Microbiology, 82*(6), 1838–1845. https://doi.org/10.1128/AEM.03163-15
Ishii, K., Fujitani, H., Soh, K., Nakagawa, T., Takahashi, R., & Tsuneda, S. (2017). Enrichment and physiological characterization of a cold-adapted nitrite-oxidizing Nitrotoga sp. from an eelgrass sediment. *Applied and Environmental Microbiology, 83*(14), e00549–17. https://doi.org/10.1128/AEM.00549-17
Ishii, K., Fujitani, H., Sekiguchi, Y., & Tsuneda, S. (2020). Physiological and genomic characterization of a new 'Candidatus Nitrotoga' isolate. *Environmental Microbiology, 22*(6), 2365–2382. https://doi.org/10.1111/1462-2920.15015
Jacob, J., Nowka, B., Merten, V., Sanders, T., Spieck, E., & Dähnke, K. (2017). Oxidation kinetics and inverse isotope effect of marine nitrite-oxidizing isolates. *Aquatic Microbial Ecology, 80*(3), 289–300. https://doi.org/10.3354/ame01859
Keuter, S., Kruse, M., Lipski, A., & Spieck, E. (2011). Relevance of Nitrospira for nitrite oxidation in a marine recirculation aquaculture system and physiological features of a Nitrospira marina-like isolate. *Environmental Microbiology, 13*(9), 2536–2547. https://doi.org/10.1111/j.1462-2920.2011.02525.x
Kitzinger, K., Koch, H., Lücker, S., Sedlacek, C. J., Herbold, C., Schwarz, J., & Daims, H. (2018). Characterization of the first "Candidatus Nitrotoga" isolate reveals metabolic versatility and separate evolution of widespread nitrite-oxidizing bacteria. *Mbio, 9*(4), e01186-e1218. https://doi.org/10.1128/mBio.01186-18
Koch, H., Lücker, S., Albertsen, M., Kitzinger, K., Herbold, C., Spieck, E., & Daims, H. (2015). Expanded metabolic versatility of ubiquitous nitrite-oxidizing bacteria from the genus Nitrospira. *Proceedings of the National Academy of Sciences, 112*(36), 11371–11376. https://doi.org/10.1073/pnas.1506533112
Lantz, M. A., Boddicker, A. M., Kain, M. P., Berg, O. M., Wham, C. D., & Mosier, A. C. (2021). Physiology of the nitrite-oxidizing bacterium Candidatus Nitrotoga sp. CP45 enriched from a Colorado river. *Frontiers in Microbiology, 12*. https://doi.org/10.3389/fmicb.2021.709371
Lebedeva, E. V., Alawi, M., Maixner, F., Jozsa, P. G., Daims, H., & Spieck, E. (2008). Physiological and phylogenetic characterization of a novel lithoautotrophic nitrite-oxidizing bacterium, 'Candidatus Nitrospira bockiana.' *International Journal of Systematic and Evolutionary Microbiology, 58*(1), 242–250. https://doi.org/10.1099/ijs.0.65379-0
Liu, H., Li, J., Zhao, Y., Xie, K., Tang, X., Wang, S., & Li, Y. (2018). Ammonia oxidizers and nitrite-oxidizing bacteria respond differently to long-term manure application in four paddy soils of south of China. *Science of the Total Environment, 633*, 641–648. https://doi.org/10.1016/j.scitotenv.2018.03.108

Lücker, S., Nowka, B., Rattei, T., Spieck, E., & Daims, H. (2013). The genome of Nitrospina gracilis illuminates the metabolism and evolution of the major marine nitrite oxidizer. *Frontiers in Microbiology, 4*, 27. https://doi.org/10.3389/fmicb.2013.00027

Ma, B., Bao, P., Wei, Y., Zhu, G., Yuan, Z., & Peng, Y. (2015). Suppressing nitrite-oxidizing bacteria growth to achieve nitrogen removal from domestic wastewater via anammox using intermittent aeration with low dissolved oxygen. *Scientific Reports, 5*(1), 1–9. https://doi.org/10.1038/srep13048

Mehrani, M. J., Sobotka, D., Kowal, P., Ciesielski, S., & Makinia, J. (2020). The occurrence and role of Nitrospira in nitrogen removal systems. *Bioresource Technology, 303*, 122936. https://doi.org/10.1016/j.biortech.2020.122936

Mellbye, B. L., Bottomley, P. J., & Sayavedra-Soto, L. A. (2015). Nitrite-oxidizing bacterium Nitrobacter winogradskyi produces N-acyl-homoserine lactone autoinducers. *Applied and Environmental Microbiology, 81*(17), 5917–5926. https://doi.org/10.1128/AEM.01103-15

Mueller, A. J., Jung, M. Y., Strachan, C. R., Herbold, C. W., Kirkegaard, R. H., Wagner, M., & Daims, H. (2021). Genomic and kinetic analysis of novel Nitrospinae enriched by cell sorting. *The ISME Journal, 15*(3), 732–745. https://doi.org/10.1038/s41396-020-00809-6

Mundinger, A. B., Lawson, C. E., Jetten, M. S., Koch, H., & Lücker, S. (2019). Cultivation and transcriptional analysis of a canonical Nitrospira under stable growth conditions. *Frontiers in Microbiology, 10*, 1325. https://doi.org/10.3389/fmicb.2019.01325

Ngugi, D. K., Blom, J., Stepanauskas, R., & Stingl, U. (2016). Diversification and niche adaptations of Nitrospina-like bacteria in the polyextreme interfaces of Red Sea brines. *The ISME Journal, 10*(6), 1383–1399. https://doi.org/10.1038/ismej.2015.214

Nowka, B., Daims, H., & Spieck, E. (2015). Comparison of oxidation kinetics of nitrite-oxidizing bacteria: Nitrite availability as a key factor in niche differentiation. *Applied and Environmental Microbiology, 81*(2), 745–753. https://doi.org/10.1128/AEM.02734-14

Off, S., Alawi, M., & Spieck, E. (2010). Enrichment and physiological characterization of a novel Nitrospira-like bacterium obtained from a marine sponge. *Applied and Environmental Microbiology, 76*(14), 4640–4646. https://doi.org/10.1128/AEM.00320-10

Park, S. J., Andrei, A. Ş, Bulzu, P. A., Kavagutti, V. S., Ghai, R., & Mosier, A. C. (2020). Expanded diversity and metabolic versatility of marine nitrite-oxidizing bacteria revealed by cultivation- and genomics-based approaches. *Applied and Environmental Microbiology, 86*(22), e01667-e1720. https://doi.org/10.1128/AEM.01667-20

Pester, M., Maixner, F., Berry, D., Rattei, T., Koch, H., Lücker, S., & Daims, H. (2014). NxrB encoding the beta subunit of nitrite oxidoreductase as functional and phylogenetic marker for nitrite-oxidizing Nitrospira. *Environmental Microbiology, 16*(10), 3055–3071. https://doi.org/10.1111/1462-2920.12300

Ruiz-Martínez, A., Claros, J., Serralta, J., Bouzas, A., & Ferrer, J. (2018). Modeling the decay of nitrite oxidizing bacteria under different reduction potential conditions. *Process Biochemistry, 71*, 159–165. https://doi.org/10.1016/j.procbio.2018.05.021

Shah, M. P. (2020). *Microbial bioremediation and biodegradation*. Springer.

Shah, M. P. (2021). *Removal of refractory pollutants from wastewater treatment plants*. CRC Press.

Sorokin, D. Y., Lücker, S., Vejmelkova, D., Kostrikina, N. A., Kleerebezem, R., Rijpstra, W. I. C., & Daims, H. (2012). Nitrification expanded: Discovery, physiology and genomics of a nitrite-oxidizing bacterium from the phylum Chloroflexi. *The ISME Journal, 6*(12), 2245–2256. https://doi.org/10.1038/ismej.2012.70

Sorokin, D. Y., Vejmelkova, D., Lücker, S., Streshinskaya, G. M., Rijpstra, W. I. C., Damste, J. S. S., & Daims, H. (2014). Nitrolancea hollandica gen. nov., sp. nov., a chemolithoautotrophic nitrite-oxidizing bacterium isolated from a bioreactor belonging to the phylum Chloroflexi. *International Journal of Systematic and Evolutionary Microbiology, 64*(Pt_6), 1859–1865. https://doi.org/10.1099/ijs.0.062232-0

Sorokin, D. Y., Lücker, S., & Daims, H. (2018). "Nitrolancea", In W. B. Whitman, F. Rainey, P. Kämpfer, M. Trujillo, J. Chun, P. DeVos et al. (Eds.), *Bergey's manual of systematics of Archaea and Bacteria*. Wiley, pp. 1–6. https://doi.org/10.1002/9781118960608.gbm01563

Spieck E., & Bock E. (2005). The lithoautotrophic nitrite-oxidizing bacteria. In D. J. Brenner, N. R. Krieg, J. T. Staley, & G. M. Garrity (Eds.), *Bergey's manual® of systematic bacteriology*. Springer. https://doi.org/10.1007/0-387-28021-9_19

Spieck, E., & Lipski, A. (2011). Cultivation, growth physiology, and chemotaxonomy of nitrite-oxidizing bacteria. *Methods in Enzymology, 486*, 109–130. https://doi.org/10.1016/B978-0-12-381294-0.00005-5

Spieck, E., Keuter, S., Wenzel, T., Bock, E., & Ludwig, W. (2014). Characterization of a new marine nitrite oxidizing bacterium, Nitrospina watsonii sp. nov., a member of the newly proposed phylum "Nitrospinae". *Systematic and Applied Microbiology, 37*(3), 170–176. https://doi.org/10.1016/j.syapm.2013.12.005

Spieck, E., Sass, K., Keuter, S., Hirschmann, S., Spohn, M., Indenbirken, D., & Giaveno, A. (2020). Defining culture conditions for the hidden nitrite-oxidizing bacterium Nitrolancea. *Frontiers in Microbiology, 11*, 1522. https://doi.org/10.3389/fmicb.2020.01522

Spieck, E., Wegen, S., & Keuter, S. (2021). Relevance of Candidatus Nitrotoga for nitrite oxidation in technical nitrogen removal systems. *Applied Microbiology and Biotechnology, 105*(19), 7123–7139. https://doi.org/10.1007/s00253-021-11487-5

Starkenburg, S. R., Chain, P. S., Sayavedra-Soto, L. A., Hauser, L., Land, M. L., Larimer, F. W., & Hickey, W. J. (2006). Genome sequence of the chemolithoautotrophic nitrite-oxidizing bacterium Nitrobacter winogradskyi Nb-255. *Applied and Environmental Microbiology, 72*(3), 2050–2063. https://doi.org/10.1128/AEM.72.3.2050-2063.2006

Starkenburg, S. R., Larimer, F. W., Stein, L. Y., Klotz, M. G., Chain, P. S., Sayavedra-Soto, L. A., & Bottomley, P. J. (2008). Complete genome sequence of Nitrobacter hamburgensis X14 and comparative genomic analysis of species within the genus Nitrobacter. *Applied and Environmental Microbiology, 74*(9), 2852–2863. https://doi.org/10.1128/AEM.02311-07

Starkenburg, S. R., Spieck, E., & Bottomley, P. J. (2011). Metabolism and genomics of nitrite-oxidizing bacteria: emphasis on studies of pure cultures and of Nitrobacter species. *Nitrification*, 265–293. https://doi.org/10.1128/9781555817145.ch11

Sun, X., Kop, L. F., Lau, M. C., Frank, J., Jayakumar, A., Lücker, S., & Ward, B. B. (2019). Uncultured Nitrospina-like species are major nitrite oxidizing bacteria in oxygen minimum zones. *The ISME Journal, 13*(10), 2391–2402. https://doi.org/10.1038/s41396-019-0443-7

Ushiki, N., Fujitani, H., Aoi, Y., & Tsuneda, S. (2013). Isolation of Nitrospira belonging to sublineage II from a wastewater treatment plant. *Microbes and Environments*, ME13042. https://doi.org/10.1264/jsme2.ME13042

Yepsen, D. V., Levipan, H. A., & Molina, V. (2019). Nitrospina bacteria in a rocky intertidal habitat (Quintay Bay, central Chile). *Microbiology Open, 8*(3), e00646. https://doi.org/10.1002/mbo3.646

Yao, Q., & Peng, D. C. (2017). Nitrite oxidizing bacteria (NOB) dominating in nitrifying community in full-scale biological nutrient removal wastewater treatment plants. *AMB Express, 7*(1), 1–11. https://doi.org/10.1186/s13568-017-0328-y

Zhang, Y., Zhang, Y., Gao, J., Shen, Q., Bai, Z., Zhuang, X., & Zhuang, G. (2018). Optimization of the medium for the growth of Nitrobacter winogradskyi by statistical method. *Letters in Applied Microbiology, 67*(3), 306–313. https://doi.org/10.1111/lam.13036

Current Perspectives of Anammox-Denitrification Technology and Its Application in Industrial Wastewater Treatment

Ravi Kumar Sonwani

Abstract A wide range of nitrogenous compounds (i.e., ammonia, nitrite, nitrate, etc.) is discharged by industrial and agricultural sectors into natural water bodies. The excess amount of nitrogenous compounds in water bodies leads to eutrophication (i.e., excess algal growth), causing the continuous deterioration of water quality. Therefore, a cost-effective and environmentally sound technique is required against such pollutants. In this direction, anaerobic ammonium oxidation-denitrification (Anammox-denitrification) technology has received significant attention to biologically converting the nitrogenous compounds into nitrogen in engineered systems. This chapter introduces a brief overview of recent advancements in Anammox-denitrification technology along with its merits and demerits. In addition, the application of Anammox-denitrification technology in industrial wastewater treatment is summarized. The kinetics of nitrogen removal by Anammox-denitrification are also discussed.

Keywords Anammox-denitrification · Kinetics · Industrial wastewater · Nitrogenous compounds · Bioreactor

1 Introduction

The discharge of a significant amount of nitrogenous compounds (i.e., ammonia, nitrite, nitrate, etc.) into the aquatic systems leads to acidification and eutrophication (i.e., excess algal growth) (Iannaconea et al., 2019; Lan et al., 2011). At the same time, the presence of an excess amount of nitrogenous compounds can also adversely affect the survival of aquatic plants and other organisms, and subsequently deteriorate the water quality (Swain et al., 2020). The main source of nitrogenous compounds in wastewater is runoff from agriculture, households, and industries (Iannacone et al.,

R. K. Sonwani (✉)
Department of Chemical Engineering, Indian Institute of Petroleum and Energy (IIPE), Visakhapatnam, Andhra Pradesh 530003, India
e-mail: ravikumar.che@iipe.ac.in

M. P. Shah (ed.), *Anammox Technology in Industrial Wastewater Treatment*,
https://doi.org/10.1007/978-981-99-3459-1_6

2019). A cost-effective and eco-friendly method is required to treat the nutrient-rich wastewater before being disposed of in open-water bodies. In this context, the conventional and still most common method is the biological process (i.e., sequential nitrification and denitrification) which is widely employed to remove nitrogenous compounds (e.g., ammonia, nitrite, nitrate, etc.) from wastewater (McCarty, 2018). According to Lan et al. (2011), sequential nitrification and denitrification is considered the highly effective and appropriate method to treat wastewater with high ammonium and carbon.

In nitrification, ammonia is oxidized to nitrite (NO_2^-) followed by nitrate (NO_3^-) under aerobic conditions. The nitrate is further reduced to gaseous nitrogen (N_2) under anaerobic conditions (Thorat et al., 2022). However, this process is energy-intensive because it requires a high oxygen supply and organic matter (e.g., methanol, ethanol, glucose, etc.) as electron donors for denitrification (Chai et al., 2019). In addition, municipal wastewater is generally contained a low carbon-to-nitrogen ratio (C/N) and subsequently leads to low nitrogen removal efficiency (Zhu et al., 2015). Due to the high energy cost and need for organic carbon in conventional nitrification and denitrification, other possible nitrogen removal processes have developed. Simultaneous anaerobic ammonium oxidation (Anammox) and Denitrification (called SAD) is an emerging biological process to effectively remove the nitrogenous compounds as well as organic pollutants in a single reactor (Lie et al., 2019; Kumar & Lin, 2010). This process is associated with several merits such as a small reactor footprint, utilizing 22–40% less carbon source, and low fabrication and energy cost than the conventional nitrification and denitrification (Bhattacharya & Mazumder, 2021; Seifi & Fazaelipoor, 2012). The microorganisms such as *Thiosphaera pantotroph*, *Pseudmonas stutzeri*, *Alcaligenes faecalis*, and *Bacillus subtilis* have been employed to treat nitrogenous compounds (Lei et al., 2019). In the last few decades, the bioreactors, namely, sequencing batch biofilm reactor (SBBR), rotating biological contactor (RBC), fluidized bed biofilm reactor (FBBR), packed bed biofilm reactor (PBBR), moving bed biofilm reactor (MBBR), and membrane biofilm reactor (MBR) have been extensively employed to achieve nearly complete removal of nitrogen (Bhattacharya & Mazumder, 2021; Hocaoglu et al., 2011).

2 Mechanisms of Simultaneous Anammox-Denitrification

Anammox-denitrification is a biological process to simultaneously remove the nitrogenous (i.e., ammonia, nitrite, nitrate, etc.) and organic compounds in engineered systems. In this method, ammonia acts as an electron donor and reacts with nitrite as the electron acceptor to form gaseous nitrogen. According to Wang et al. (2021), nitrite (NO_2^-) instead of nitrate (NO_3^-) was the direct oxidized agent for ammonia, and therefore, denitrification plays a significant role in providing nitrate for Anammox by reducing nitrite. The mechanisms of simultaneous Anammox-denitrification can broadly be classified into three types (Wang et al., 2021; You et al., 2020). (i) In the first type of Anammox/denitratation process ($NO_3^- \rightarrow NO_2^-$),

Anammox obtains NO_2^- through denitratation with NO_3^- act as the substrate (Fig. 1a). Along with partial nitrification, denitratation is also recognized as an effective route to provide NO_2^- for Anammox. (ii) In second type of Anammox/denitratation ($NO_2^- \rightarrow N_2$) process, Anammox and denitritation simultaneously contribute to convert the NO_2^- into N_2, particularly when NO_2^- is the main substrate than NH_4^+ (Fig. 1b). (iii) In third type Anammox/(denitratation + denitritation) ($NO_3^- \rightarrow$ N2), NO_3^- and NO_2^- are removed by denitratation + denitritation like a feedback loop (Fig. 1c, d) (Wang et al., 2021).

3 Parameters Affecting the Performance of Anammox-Denitrification

The efficiency of nitrogenous compounds removal in Anammox-denitrification process is mainly affected by the parameters, namely, carbon to nitrogen ratio (C/N), pH, temperature, dissolved oxygen, ammonia, and nitrite concentration (Bhattacharya & Mazumder, 2021; You et al., 2020). Some of these parameters are discussed below.

3.1 PH

pH is the key parameter that influences the overall efficacy of the process. The concentration of ammonia and nitrous acid varies with the pH of wastewater. At low pH, the amount of free nitrous acid generally increases, while, at high pH, the amount of free ammonia rises (You et al., 2020). Generally, the consumption of hydrogen ion (H^+), subsequently the pH in Anammox and partial denitrification is increased (Kumar & Lin, 2010). The yield and denitrifying rates decrease under acidic conditions ($pH < 5$) which could be due to the formation of nitrous oxide (NO) (Kumar & Lin, 2010). If the pH of the solution is below 5.0 or above 9.0, the growth and reproduction bacteria are adversely affected which consequently inhibits the overall performance. The optimum pH range for the Anammox and denitrification is 6.7–9.5 and 6.0–9.0, respectively (Tomaszewski et al., 2017; Kumar & Lin, 2010).

3.2 Temperature

Temperature is also a significant constraint that directly influences the growth of microorganisms and oxygen solubility (Sonwani et al., 2021a). It also affects the rate of dissolved oxygen and settling characteristics of biological solids. Anammox bacteria can survive between 20 °C and 40 °C, whereas denitrifying bacteria can also

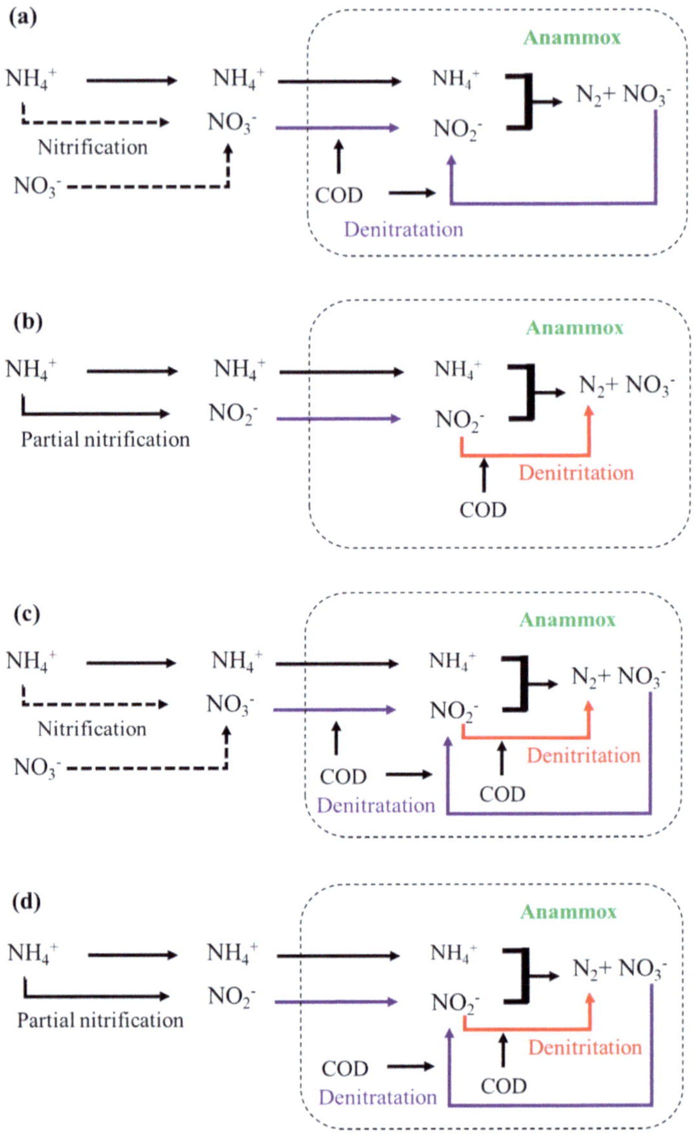

Fig. 1 Simultaneous Anammox-denitrification processes. **a** Anammox/denitratation; **b** anammox/denitritation; **c** Anammox/(denitratation + denitritation) with NO_3^- as the major electron acceptor, and **d** Anammox/(denitratation + denitritation with NO_2^- as the major electron acceptor (Reproduced from Wang et al., 2021)

effectively work between 20 °C and 35 °C (Kumar & Lin, 2010). The ideal temperature range for the Anammox process is 30 °C–40 °C. In the Anammox process, the temperature beyond 45 °C inhibits the activity of Anammox bacteria because the enzyme may get denatured at high temperature. However, it has also been reported that the metabolic activity of Anammox bacteria gets reduced at low temperature (<10 °C). Almost similar kind of temperature effect is generally observed in the denitrification process. Based on the previous studies, 30 °C–35 °C is optimal for simultaneous Anammox and partial denitrification process to achieve the maximum removal of nitrogenous compounds (You et al., 2020).

3.3 Carbon to Nitrogen Ratio (C/N)

Carbon to nitrogen ratio (C/N) is also important for microbial growth because carbon is required as an energy source, whereas nitrogen facilitates the synthesis of amino acids, proteins, and nucleic acids (Xi et al., 2022). It also helps to keep the bioreactor (such as sequencing batch reactor, moving bed bioreactor, membrane bioreactor, etc.) stable during operation (You et al., 2020). The high carbon exerts a negative impact as heterotrophic denitrifying bacteria cultivate at a much higher rate than the autotrophic Anammox bacteria (Kumar & Lin, 2010). The presence of a high carbon inhibits Anammox bacterial growth. Various researchers have reported that the optimum C/N ratio in the combined Anammox and denitrification process is varied between 2.0 and 3.0 for effective startup and operation of the bioreactors (Cao et al., 2019; You et al., 2020). According to Wang et al. (2021), the wastewater has a low C/N ratio exerted more than 60% removal of total nitrogen in bioreactor.

3.4 Dissolved Oxygen (DO) and Oxidation–Reduction Potential (ORP)

DO and ORP parameters affect the overall efficiency of the Anammox and denitrification process. It has been reported that the lower DO (less than 2% of air saturation) impedes the metabolic activity of Anammox bacteria. In addition, very high DO (more than 18% of air saturation) irreversibly inhibits the activity of bacteria. Anammox bacteria generally do not survive at very high DO (Yin et al., 2016). Also, denitrifying bacteria survive effectively at close to zero DO. Therefore, simultaneous Anammox and some denitrification bacteria require very low DO to survive, generally below 0.7 mg/L (You et al., 2020). ORP is employed to determine the macroscopic redox properties shown by matters in an aqueous solution and can be correlated to the parameters namely, DO, pH, and organic matter loading. The high value of ORP indicates the presence of high DO in wastewater. An ORP between -50 and $+ 50$ mV is most suitable for simultaneous Anammox and denitrification

(You et al., 2020). ERP is extensively employed to monitor and control the Anammox and denitrification process (Yang et al., 2015).

4 Role of Kinetic Models

The kinetic models play a key role to evaluate nitrogen removal kinetics and bioreactor designing. The models such as the first-order model, Grau second-order model, and Monod model are widely employed to estimate the nitrogen removal kinetics (Ni et al., 2010; Sonwani et al., 2021b).

4.1 *First-Order Kinetic Model*

The change of substrate concentration in the complete mixed system can be represented as (Swain et al., 2021).

$$-\frac{dS}{dt} = \frac{QS_i}{V} - \frac{Q.S_e}{V} - k_1 S_e \tag{1}$$

Under the pseudo-steady-state condition, the change of substrate (*dS/dt*) within the bioreactor becomes insignificant, and Eq. (1) can be written as;

$$\frac{(S_i - S_e)}{HRT} = k_1.S_e \tag{2}$$

where S_i and S_e are the initial and effluent total nitrogen concentrations (mg/L), HRT is the hydraulic retention time (day), k_1 represents the first-order rate constant (day^{-1}), Q is the volumetric inlet flow rate (L/d), and V is the working reactor volume (L).

4.2 *Grau Second-Order Model*

The general equation of a second-order model is represented below (Sonwani et al., 2021a).

$$-\frac{dS}{dt} = \frac{Q.S_i}{V} - \frac{Q.S_e}{V} - k_2 X\left(\frac{S_e}{Si}\right)^2 \tag{3}$$

By integration and linearization, the above equation can be stated as follows:

$$\frac{S_i.HRT}{(S_i - S_e)} = HRT + \frac{S_i}{k_{2X}} \tag{4}$$

where $S_i/(k_2.X)$ is considered as constant. The term $(S_i\text{-}S_e/S_i)$ is the substrate removal efficiency. The above equation can be represented as.

$$\frac{HRT}{E} = b.HRT + a \tag{5}$$

where a and b are the Grau kinetic constants. E is the total nitrogen removal efficiency (%), k_2 represents the Grau second-order rate constant (d^{-1}), and X is the bacterial biomass concentration (mg/L).

4.3 Monod Model

Monod model is a well-known expression to describe microbial growth (Sonwani et al., 2021a) as given in the following equation.

$$\mu = \frac{\mu_{max} S}{K_s + S} \tag{6}$$

where μ, μ_{max}, K_s, and S represent the specific growth rate of microbes (d^{-1}), the maximum specific growth rate of microbes (d^{-1}), half-saturation rate constant (mg/L), and substrate concentration (mg/L), respectively. These constants (Ks, μ, and μ_{max}) are very important to estimate the microbial growth rate in the bioreactor. The microbial growth rate is generally estimated by the following expression.

$$\frac{dX}{dt} = r_g = \mu X \tag{7}$$

where r_g is the growth rate of biomass (g/L·d).

5 Application of Anammox-Denitrification in Industrial Wastewater Treatment

In the last few years, the presence of nitrogenous compounds in water bodies has become a major environmental concern due to their adverse impacts on aquatic life and human health. Therefore, nitrogenous compounds must be removed from wastewater before release into the environment. In this context, the Anammox-denitrification process has been extensively employed for the treatment of nitrogenous compounds (i.e., ammonia, nitrite, nitrate, etc.) from wastewater. An Anammox-denitrification-based sequencing batch reactor (SBR) was applied to treat the wastewater containing NH_4^+–N, (NH_4Cl), NO_2–N, ($NaNO_2$) (Du et al., 2015). The optimum nitrogen removal efficiency of 94% was achieved with 10.1 mg/L of initial total nitrogen in wastewater. In another study, a SBR was employed to treat the wastewater comprising nitrate (100 mg–N/L), ammonium (70 mg–N/L), and acetate (50–250 mg–C/L) (Takekawa et al., 2014). SBR effectively removed 58–94% of total nitrogen under a low C/N ratio (0.5–1.0) and 67–79% of total nitrogen under a high C/N ratio (1.2–2.5). An upflow anaerobic sludge blanket bioreactor (UASBR) was used to remove total nitrogen based on a simultaneous Anammox and denitrification process (Wang et al., 2019). The maximum nitrogen removal efficiency of 92.47% was obtained at 0.9 kg–N/m^3 day of total nitrogen. *Thauera* and *Candidatus Brocadia* were the dominant microorganisms to remove the total nitrogen from wastewater.

In the presence of 150 mg/L of COD as carbon source, a SBR (based on simultaneous Anammox and denitrification) completely removed ammonia, nitrite, and nitrate from wastewater (Li et al., 2016). The synergy of Anammox and denitrification simultaneously removed carbon and nitrogen. They also reported that Anammox bacteria were leading at low COD, whereas denitrifying bacteria predominated at high COD. Wang and He (2022), employed Anammox enriched culture (i.e., AMX) for the complete removal of nitrogenous compounds. They found that Anammox and denitrifying bacteria involving soluble microbial products (SMPs) consumption and poly-β-hydroxybutyrate (PHB) synthesis played a significant role in nitrogen removal. The efficacy of two SBRs added with carbons, namely, acetate (R1) and ethanol (R2) was evaluated for simultaneously treating NH_4^+–N and NO_3^-N from wastewater (Du et al., 2017). The nitrogen removal efficiency of 93.6% was achieved in R1 with influent nitrate to ammonia ratio of 1.0, whereas lower nitrogen removal efficiency (90.0%) was obtained in R2. The microorganisms, namely, *Thauera* genera, *Candidatus Brocadia*, and *Candidatus Kuenenia* were played a significant role in nitrogen removal.

6 Conclusion

Excess nitrogenous compounds into aquatic systems exert adverse impacts on the ecosystem as well as deteriorate the water quality. Among various treatment technologies, simultaneous Anammox-denitrification holds significant attention to biologically convert the nitrogenous compounds into diatomic nitrogen. This chapter delivered an outline of the present state of research and development of Anammox-denitrification technology. The parameters, namely, C/N, pH, temperature, DO, and ORP are important for the stable bioreactor operation and subsequently achieve the optimum nitrogen removal rate. The models, namely, the first-order model, Grau second-order model, and Monod model are supportive to estimate the nitrogen removal kinetics and bioreactor designing. In order to improve the efficacy of the Anammox-denitrification process, studies on the mass transfer mechanism and kinetic model for coupling inhibition need to be further explored.

Acknowledgements The author thankfully acknowledges the Indian Institute of Petroleum and Energy (IIPE), Visakhapatnam, India for providing financial support and laboratory facilities to execute the work.

References

Bhattacharya, R., & Mazumder, D. (2021). Simultaneous nitrification and denitrification in moving bed bioreactor and other biological systems. *Bioprocess and Biosystems Engineering, 44*(4), 635–652.

Cao, S., Du, R., Peng, Y., Li, B., & Wang, S. (2019). Novel two stage partial denitrification (PD)-Anammox process for tertiary nitrogen removal from low carbon/nitrogen (C/N) municipal sewage. *Chemical Engineering Journal, 362*, 107–115.

Chai, H., Xiang, Y., Chen, R., Shao, Z., Gu, L., Li, L., & He, Q. (2019). Enhanced simultaneous nitrification and denitrification in treating low carbon-to-nitrogen ratio wastewater: Treatment performance and nitrogen removal pathway. *Bioresource Technology, 280*, 51–58.

Du, R., Peng, Y., Cao, S., Wang, S., & Wu, C. (2015). Advanced nitrogen removal from wastewater by combining anammox with partial denitrification. *Bioresource Technology, 179*, 497–504.

Du, R., Cao, S., Li, B., Niu, M., Wang, S., & Peng, Y. (2017). Performance and microbial community analysis of a novel DEAMOX based on partial-denitrification and anammox treating ammonia and nitrate wastewaters. *Water Research, 108*, 46–56.

Hocaoglu, S. M., Insel, G., Cokgor, E. U., & Orhon, D. (2011). Effect of sludge age on simultaneous nitrification and denitrification in membrane bioreactor. *Bioresource Technology, 102*(12), 6665–6672.

Iannacone, F., Di Capua, F., Granata, F., Gargano, R., Pirozzi, F., & Esposito, G. (2019). Effect of carbon-to-nitrogen ratio on simultaneous nitrification denitrification and phosphorus removal in a microaerobic moving bed biofilm reactor. *Journal of Environmental Management, 250*, 109518.

Kumar, M., & Lin, J. G. (2010). Co-existence of anammox and denitrification for simultaneous nitrogen and carbon removal—strategies and issues. *Journal of Hazardous Materials, 178*(1–3), 1–9.

Lan, C. J., Kumar, M., Wang, C. C., & Lin, J. G. (2011). Development of simultaneous partial nitrification, anammox and denitrification (SNAD) process in a sequential batch reactor. *Bioresource Technology, 102*(9), 5514–5519.

Lei, X., Jia, Y., Chen, Y., & Hu, Y. (2019). Simultaneous nitrification and denitrification without nitrite accumulation by a novel isolated Ochrobactrum anthropic LJ81. *Bioresource Technology, 272*, 442–450.

Li, J., Qiang, Z., Yu, D., Wang, D., Zhang, P., & Li, Y. (2016). Performance and microbial community of simultaneous anammox and denitrification (SAD) process in a sequencing batch reactor. *Bioresource Technology, 218*, 1064–1072.

McCarty, P. L. (2018). What is the best biological process for nitrogen removal: When and why? *Environmental Science and Technology, 52*, 3835–3841.

Ni, S. Q., Lee, P. H., & Sung, S. (2010). The kinetics of nitrogen removal and biogas production in an anammox non-woven membrane reactor. *Bioresource Technology, 101*(15), 5767–5773.

Seifi, M., & Fazaelipoor, M. H. (2012). Modeling simultaneous nitrification and denitrification (SND) in a fluidized bed biofilm reactor. *Applied Mathematical Modelling, 36*(11), 5603–5613.

Sonwani, R. K., Kim, K. H., Zhang, M., Tsang, Y. F., Lee, S. S., Giri, B. S., Singh, R. S., & Rai, B. N. (2021a). Construction of biotreatment platforms for aromatic hydrocarbons and their future perspectives. *Journal of Hazardous Materials, 416*, 125968.

Sonwani, R. K., Swain, G., Jaiswal, R. P., Singh, R. S., & Rai, B. N. (2021b). Moving bed biofilm reactor with immobilized low-density polyethylene–polypropylene for Congo red dye removal. *Environmental Technology and Innovation, 23*, 101558.

Swain, G., Sonwani, R. K., Giri, B. S., Singh, R. S., Jaiswal, R. P., & Rai, B. N. (2020). Collective removal of phenol and ammonia in a moving bed biofilm reactor using modified bio-carriers: Process optimization and kinetic study. *Bioresource Technology, 306*, 123177.

Swain, G., Singh, S., Sonwani, R. K., Singh, R. S., Jaiswal, R. P., & Rai, B. N. (2021). Removal of Acid Orange 7 dye in a packed bed bioreactor: Process optimization using response surface methodology and kinetic study. *Bioresource Technology Reports, 13*, 100620.

Takekawa, M., Park, G., Soda, S., & Ike, M. (2014). Simultaneous anammox and denitrification (SAD) process in sequencing batch reactors. *Bioresource Technology, 174*, 159–166.

Thorat, B. N., & Sonwani, R. K. (2022). Current technologies and future perspectives for the treatment of complex petroleum refinery wastewater: A review. *Bioresource Technology*, 127263.

Tomaszewski, M., Cema, G., & Ziembińska-Buczyńska, A. (2017). Influence of temperature and pH on the anammox process: A review and meta-analysis. *Chemosphere, 182*, 203–214.

Wang, Q., & He, J. (2022). Partnering of anammox and denitrifying bacteria benefits anammox's recovery from starvation and complete nitrogen removal. *Science of the Total Environment, 815*, 152696.

Wang, D., He, Y., & Zhang, X. X. (2019). A comprehensive insight into the functional bacteria and genes and their roles in simultaneous denitrification and anammox system at varying substrate loadings. *Applied Microbiology and Biotechnology, 103*(3), 1523–1533.

Wang, B., Gong, X., & Peng, Y. (2021). Simultaneous anammox-denitrification process and its emerging extensions. *Chemical Engineering Journal, 415*, 128380.

Xi, H., Zhou, X., Arslan, M., Luo, Z., Wei, J., Wu, Z., & El-Din, M. G. (2022). Heterotrophic nitrification and aerobic denitrification process: Promising but a long way to go in the wastewater treatment. *Science of the Total Environment, 805*, 150212.

Yang, J., Trela, J., Plaza, E., Wahlberg, O., & Levlin, E. (2015). Oxidation-reduction potential (ORP) as a control parameter in a single-stage partial nitritation/anammox process treating reject water. *Journal of Chemical Technology and Biotechnology, 91*(10), 2582–2589.

Yin, Z., dos Santos, C. E. D., Vilaplana, J. G., Sobotka, D., Czerwionka, K., Damianovic, M. H. R. Z., & Makinia, J. (2016). Importance of the combined effects of dissolved oxygen and pH on optimization of nitrogen removal in anammox-enriched granular sludge. *Process Biochemistry, 51*(9), 1274–1282.

You, Q. G., Wang, J. H., Qi, G. X., Zhou, Y. M., Guo, Z. W., Shen, Y., & Gao, X. (2020). Anammox and partial denitrification coupling: A review. *Rsc Advances, 10*(21), 12554–12572.

Zhu, T., Zhang, Y., Quan, X., & Li, H. (2015). Effects of an electric field and iron electrode on anaerobic denitrification at low C/N ratios. *Chemical Engineering Journal, 266*, 241–248.

Isolation, Physiological Characteristics, Ecological Importance, and Chemotaxonomy of Nitrite-Oxidizing Bacteria with Their Associated Genes in Nitrogen Fixation

Ritu Rani, Jitender Rathee, Nater Pal Singh, and Anita Rani Santal

Abstract Nitrite-oxidizing bacteria (NOB) are primarily organo and/or chemoautotrophic Gram-negative bacteria that play a critical role in aerobic nitrification and gain energy by converting nitrite into nitrate. NOB is found in natural ecosystems and also in man-made ecosystems. Studies show that NOB lives in extreme environmental conditions such as concrete and natural stones, acidic soil, alkaline soil, and hot water springs. Some species can also survive in a long period of starvation and in dry conditions. They also need a specific type of enriched media composition and laboratory conditions for their growth. Chemotaxonomy helps to characterize NOB in mixed and pure cultures based on biochemical composition. In this chapter, different methods and techniques for the isolation and characterization of NOB from different environmental conditions will be discussed.

Keywords Nitrification · Nitrite-oxidizing bacteria · Heterotrophic · Nitrite · Isolation

1 Introduction

Nitrogen is essential for the entire living organism present on earth as it is present in DNA, RNA, amino acids, protein, etc. Microorganisms play a vital role in the biogeochemical nitrogen cycle, transforming nitrogen molecule into various forms such as organic molecules, which are required for life and subsequently recycling nitrogen from decaying biomass into the atmosphere (Damis et al., 2016). Nitrification is a critical phase in the biogeochemical nitrogen cycle, converting NH_3 to nitrite and nitrate. Sergei Winogradsky was the first to produce chemolithoautotrophic microbes by the

R. Rani · J. Rathee · A. R. Santal (✉)
Department of Microbiology, Maharshi Dayanand University, Rohtak 124001, Haryana, India
e-mail: anita.gangotra@gmail.com; anita.micro@mdurohtak.ac.in

N. P. Singh
Centre for Biotechnology, Maharshi Dayanand University, Rohtak 124001, Haryana, India

M. P. Shah (ed.), *Anammox Technology in Industrial Wastewater Treatment*,
https://doi.org/10.1007/978-981-99-3459-1_7

use of NH_3 or NO_2^- as an electron donor. Winogradsky discovered that nitrification takes place in two steps and is catalyzed by two different types of bacteria: AOB for ammonia oxidation and NOB for nitrite oxidation, both of which are required to complete the nitrification process. Both the AOB and NOB groups were slow-growing bacteria (Dworkin et al., 2012). Because ammonia oxidation is the rate-limiting nitrification phase, research on ammonia oxidizers has mostly concentrated on identifying ammonia-oxidizing archaea (AOA) (Könneke et al., 2005). The mechanism of nitrite oxidation as an energy source has earlier been described. The nitrite oxidoreductase (NOR) enzyme is responsible for the importance of nitrite-oxidizing bacteria (Abeliovich, 2006). The reaction is as follows:

$$NO_2 + H_2O \rightarrow NO_3^- + 2H_2O + e^- + 2H^+$$

$$2H^+ + e^- + 0.5O_2 \rightarrow 2H_2O$$

The techniques such as PAGE and size exclusion chromatography revealed that the enzyme extracted from heat-treated membranes was homogeneous and isolated from various strains of *Nitrobacter*, including mixotrophically generated cells of *Nitrobacter hamburgensis*. Each enzyme molecule contains 0.70 g-atoms of molybdenum, 23.0 g-atoms of iron, 1.76 g-atoms of zinc, and 0.89 g-atoms of copper. The molybdenum and iron-sulfur center of Nitrobacter hamburgensis contains nitrite oxidoreductase linked to the conversion of nitrite to nitrate (Abeliovich, 2006).

Nitrogen loss in the soil occurs when ammonium fertilizer is transformed to nitrate in the agricultural industry. The nitrite oxidation mechanism must be understood in this case. By eliminating excess nitrogen from sewage water, the nitrite oxidation process aids microbiological wastewater treatment. NOB loses nitrogen as it converts nitrite to nitrate, that is used as a nitrogen source by all microbes and plants (Fig. 1) (Gruber et al., 2008). As a result, NOB plays a chief role in the biogeochemical nitrogen cycle. NOB activity was volatile in industrial wastewater treatment plants, and nitrite oxidation was not done during treatment, resulting in significant ecological harm if nitrite was released from wastewater. Despite its ecological benefits and our limited understanding of its physiology and function NOB is considered as a poorly investigated portion of the nitrogen cycle (Damis et al., 2016).

NOB is improving using metagenomic and other meta-omics methodologies, updated molecular techniques, and rectification cultivation methods, which offer a new approach to investigate NOB due to its importance in the nitrogen cycle (Albertsen et al., 2013; Shah, 2020, 2021). The majority of *Nitrobacter* species were extracted at a concentration of 29 mM nitrite, while marine strains needed a 1 mM nitrate (Table 1) (Jacob et al., 2017). However, it was later shown that different soil NOB populations have different nitrite preferences (Wang et al., 2015).

It was also observed that lowering the temperature of incubation (10 °C) was essential and NOB can be detected in wastewater treatment plants (Nowka et al., 2015). Higher growing temperature, at 60 °C, is essential to maintain NOB isolated from hot springs (Spieck & Lipski, 2011). NOB enrichments have been observed

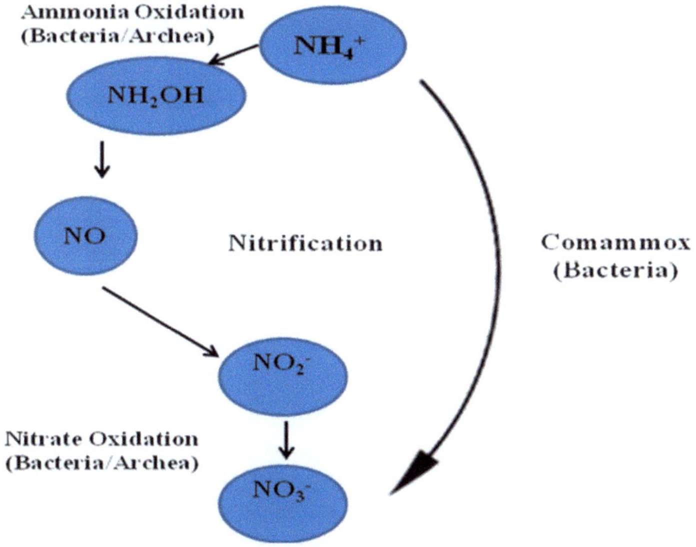

Fig. 1 Mechanism of nitrite oxidation by nitrite-oxidizing bacteria (Lancaster et al., 2018)

Table 1 Six nitrite-oxidizing bacterial genera

Serial number	Genera of Nitrite oxidizers	References
1	*Nitrobacter*	Boon and Laudelout (1962)
2	*Nitrospira*	Daims and Wagner (2018)
3	*Nitrotoga*	Lücker et al. (2015)
4	*Nitrococcus*	Watson and Waterbury (1971)
5	*Nitrospina*	Spieck and Bock (2015)
6	*Nitrolancetus*	Kruse et al. (2013)

in different environments, and it's intriguing to suppose that the complete NOB diversity is still unknown. This book chapter may help in the rediscovery of bacteria that perform the essential nitrification second step, which is often disregarded (Tao & Gao, 2012).

2 Isolation and Nutrient's Requirement of NOB

For isolation of NOB, various protocols can be used. For this purpose only sterilized mineral nitrite media could be used; otherwise, heterotrophic microorganisms will grow as nitrite oxidizers. After the isolation, isolated colonies can be extracted from the petri dishes and then tested for further to remove nitrogenous compounds (Bock & Koops, 1992).

2.1 Various Types of Media Required for the Growth of NOB

The metabolic flexibility of NOB is being used to determine the composition of diverse media. Diluted stock solutions seem to be the simplest approach to prepare media. To prevent toxicity, the sole electron donor of nitrite is usually added last, at the specific proportions recommended by the specific NOB. As lithoautotrophic growth medium, mineral salts are being used (Bock et al., 1983). A tenfold concentrated stock solution was prepared, and 100 mL of it was transferred to the Final medium, that comprised 900 mL of distilled water (pH 7.4–7.6). For smaller bottles, use plastic or metal caps, whereas for larger culture vessels use cotton plugs. For NOB cultures, specific glassware and equipment should be used beside this quality of water and a reagent also affects the growth. Media should be prepared 1–2 days prior to allow the pH to settle and become saturated with CO_2.

Plates are made with mixotrophic media containing 13 g agarose per L and $NaNO_2$ per L. (Lipski et al., 2001). In the context of *Nitrobacter*, organic material concentration can be boosted by a factor of 10. For two weeks, purity media (5 mL) loaded with one drop of NOB culture was incubated. For the preparation of marine media, tenfold concentrated stock solution was prepared for marine media, and 100 mL stock solution was added to the final medium, which contained 630 mL seawater, 270 mL distilled water, 1 mL trace elements (marine). Preheat the oven to 350°F and set the temperature to 65–70 °C.

2.2 Isolation of NOB Cultures

Extensive enrichment, as well as repeated end-point dilution, was required for successful separation of NOB. The oxidation of ammonium via nitrite to nitrate by chemolithotrophic nitrifying bacteria requires the presence of O_2. The ability of both NOB and AOB to sequester oxygen is critical to the effectiveness of nitrification in oxygen-limited situations. The culture was discovered to inhibit oxidation of nitrite prior to ammonium oxidation; hence, nitrite accumulated and collected in the body. Organotrophic bacteria compete with ammonium- and nitrite-oxidizing bacteria for available oxygen. The competition's outcome is determined by their individual oxygen affinities and population numbers. The mixotrophic proliferation of nitrite-oxidizing *Nitrobacter hamburgensis* and the AOB, compete for limited quantities of oxygen (Sliekers et al., 2005).

2.2.1 Dilution Series

To distinguish NOB from heterotrophic bacteria, a MM which contains nitrite (10 mL medium/tube) might be utilized. Nitrite intake should be monitored on a regular basis and enrichment cultures must be incubated for one to several months at successive dilutions (101–108). Continue to incubate until the number of positive tubes does not change color any more. Before transfer, all tubes were mixed at high speed using a vortex mixer to provide a nearly homogeneous distribution. Micro colonies can be manually broken by sterile glass beads to evade the associated heterotrophic bacteria, especially in the case of *Nitrospira.* Cells should be transferred from the tube with the highest dilution. Purity can be determined using a variety of complicated mediums (Park et al., 2020).

2.2.2 Plating Methods

NOB are sensitive to the amount of oxygen in the air, they do not develop on agar surfaces. Nonetheless, two species of *Nitrobacter* were isolated effectively (Sorokin et al., 2012). After many weeks of incubation on agar media, *Nitrobacter* forms small brownish colonies with a circular border on mixotrophic medium. Plates should be wrapped in parafilm to prevent drying and placed in a plastic bag to limit oxygen concentration. Instead of agar, use agarose (13 g/L) for *Nitrospira*, and irregular brownish colonies emerge after 1–3 months. Colonies are often less than a millimeter in diameter, thus they must be found under magnification before being picked up with an Eppendorf tip and inoculated into a broth medium. Because of their oxygen sensitivity, nitrite-oxidizing bacteria are thought to be unable to establish colonies (Bock et al., 1990). Despite this, the alkaline habitat isolates are able to grow slowly on the agar surface and after 14–20 days colonies would be appeared. Colonies in their early stages would be soft and pale golden in hue. Old colonies (2 months) were appeared brownish and covered in a skin-like matrix layer that likely served as an oxygen barrier (Waheed et al., 2013).

2.2.3 Density Gradient Centrifugation

Cells from a 1.5 L batch culture should be centrifuged in a little amount of sterilized 0.9% NaCl media. The cell suspension must be combined with 12 mL NaCl (1.5%) and 28 mL Percoll solution and centrifuged in 40 mL tubes for 2 h at 12,000 g and 4 °C. Each cell type will produce one brownish band in the gradient if the cell density is sufficient. Using sterile pasteur pipettes, remove the cells from the tube, resuspend in 0.9% sodium chloride, and examine under a light microscope. After 2–3 months of incubation, check for growth (Patel et al., 2022).

3 Cultivation Procedure for NOB

To initiate cultivation from ambient material, use 1 mL inoculums/150 mL media in flasks. The temperature fluctuates depending on the conditions in situ (10–46 °C) and must be carried out in the dark without movement. Replace any moisture losses in thermophilic cultures with sterilized distilled water. *Nitrospira* genus can also be grown in screw-top bottles in broth medium and have a low oxygen level (approximately 4 mg O_2/L). If a substrate with sodium nitrite concentration of 2.9 mM or less will be used, then there is a need to replace it every 7–30 days if growth occurs. The simplified Griess–Ilosvay spot test could be used to evaluate nitrite consumption. If the culture does not become pink, it is active, and the initial nitrite concentration should be replaced aseptically with a stock solution having a concentration of 2.5 M (Schmidt & Belser, 1994).

In the presence of 10 mM nitrate, delayed nitrite oxidation can be observed in a marine NOB *Nitrospira* strain. A 1–10% inoculum should be transferred every 3–12 months. In the event of contamination or other issues, maintain the old cultures from previous cultures along with the current cultures. For on-the-spot test, a modified Griess–Ilosvay reagent was used, merge two solutions and preserve them in the dim light at 4 °C.

The batch culture must be shaken after the bacterial culture has taken the substrate during the initial stage. Colonies are very sensitive to high partial pressures of O_2 initially and with stirring mechanism the cell density can be increased. Centrifugation is commonly used to obtain the cells (15,000 g). Filtration is another option; however, membrane filters might be clogged by flocs. When the cells have a strong nitrite-oxidizing activity, the pellet turns dark, but when flocs develop, it turns light. Centrifugation may result in the loss of *Candidatus Nitrotoga* or *Nitrospira* microcolonies if the aggregates do not adhere to the centrifuge bottle walls. Failure to attach/sediment primarily affected by divalent cations and hydrophobic interactions maintains the difference in surface charge between planktonic cells and microcolonies (Larsen et al., 2008). Higher *Nitrospira* microcolonies, as a result, are now more resistant to extreme shear forces and physical/chemical manipulations.

To address this issue, a novel approach of utilizing the development of gas-permeable membranes in flasks has been used (Bock et al., 1988). Nitrite and organic molecules are electron donors, and nitrate is the electron acceptor at the conclusion of the chain. Nitrite is first oxidized to nitrate, which is then reduced back to nitrite. Ammonia and nitrogenous gases (NO, N_2O) are also produced under these growth circumstances. The reactor will be run at a temperature of 28 °C. After some days, an asymmetrical (aerobic/anaerobic) biofilm will develop on the silicone membrane's surface, ensuring continuous air supply through a silicone tube. This prevents anaerobically developing cells from being inhibited by nitrite.

4 Maintenance of NOB Cultures in Lab Conditions

Nitrobacter cells can be regenerated even after many years when liquid cultures are preserved at 17 °C in the darkness. This could be challenging for more sensitive species, such as *Nitrospira*; in this instance, a high inoculum of 10% (v/v) must be used, and the substrate concentration should be lowered. The sodium nitrite concentration should be reduced to 0.15 mM at first, especially for sea strains. Freezing with liquid nitrogen can be used for maintenance. Cryoprotecting buffers, such as Hatefi, can be used to sustain stock cultures. Cultivating *Nitrobacter* in 1 L bottles filled to the full with mixotrophic media is another approach for preserving it for a long time. Instead of pyruvate, glycerol must be utilized to keep the pH stable for a long time. Through dissimilatory nitrate reduction, *Nitrobacter* has been shown to thrive anaerobically (Lücker et al., 2010). As a carbon source, external electrons from organic materials are required.

5 Monitoring the Cultures

Different NOB genera can flourish in similar enrichment culture as a result of this diversity. The most prominent cells can be recognized by phase contrast microscopy or fluorescence microscopy using DAPI labeling (Lebedeva et al., 2005). Electron microscopy can be employed to examine the existence of intercellular membrane (ICM) (Watson et al., 1989) and the periplasmic space expansions in ultrathin sections of *Nitrobacter, Nitrococcus, Nitrospira,* and *Nitrotoga,* respectively (Vijayan et al., 2021). Although spherical forms proliferate in senescent cultures, *Nitrospina* members have the appearance of long, slender rods. In a *Nitrospira* life cycle that is intriguing, planktonic cells alternate with dense microcolonies (Ushiki et al., 2013). The immunofluorescence or Western blotting techniques are employed to differentiate NOB; monoclonal antibodies targeting the NXR describe the majority of characterization of NOB using molecular techniques (PCR or FISH). Acid biomass hydrolysis allows for fatty acid analysis and chemotaxonomic profiling of NOB, which is consistent with their phylogenetic membership in distinct *Proteobacteria* subclasses or the deeper branching phylum *Nitrospirae*. Furthermore, these diverse chemotaxonomic lipid patterns aid in the identification of *Nitrospira* species (Vijayan et al., 2021).

6 Phylogeny and Genes Associated with Nitrite Oxidation

As nitrifiers grow slowly and it's very difficult to isolate, researchers have used culture-independent molecular techniques that target 16S rRNA to investigate their richness and diversity in native and controlled environments (Janssen, 2006). Without the use of particular functional markers, 16S rRNA gene sequences reveal a wide range of nitrifying groups in a microbial population (Nocker et al., 2007). A frequent ammonia oxidizer has a gene that is, *amoA* gene which codes for the alpha subunit of ammonium monooxygenase (Koper et al., 2004). Genes expressing the alpha or beta subunits of nitrite oxidoreductase (*nxrA* or *nxrB*) have been generated to detect and quantify NOB in pure cultures. Functional nitrifier genetic markers like *amoA* and *nxrB* have a greater phylogenetic resolution than 16S rRNA when it comes to distinguishing nitrifiers at the strain and species levels (Rani et al., 2017). Under organically and conventional nitrogen maintenance, *amoA* can be employed as a molecular marker to evaluate temporal changes in the abundance and diversity of AOA and AOB. The amount and community of AOB were observed to be greatly altered by 3-year repeating mineral nitrogen fertilization, but not AOA (Ouyang & Norton, 2020). There are two genes which are *nxrX* and *nxrB1* gene that play a vital role in the nitrite oxidation and these genes can be easily confirmed by 16S rRNA gene with distinguishing property. These genes form an enzyme called nitrite oxidoreductase which help in the oxidation of nitrite into nitrate compound by NOB. There is another gene that is *nxrA* gene which encodes this enzyme in the *Nitrobacter* bacteria in the soil.

7 Physiology Analysis of NOB

Pure cultures allow researchers to examine physiological aspects that are relevant to the environment. To evaluate the nitrification potential in terms of organic substances, strains growing in diverse circumstances, or additive inhibition, activity assays can be used. To determine nitrite-oxidizing activity, chemical studies are necessary, which may be done in two ways: nitrifying activity in the short term or potential nitrite-oxidizing activity. Prosser (1989) provided a great review of physiological research using nitrifying bacteria in general. A recent assessment of NOB metabolism in conjunction with a genome sequence analysis also discussed how environmental influences affect their bioenergetics and growth characteristics. The *nxr* gene with its loci has important role which carry cytoplasmic NXR (Fig. 2) and form a NXR module that gains ATP (energy) and spread it for the oxidation of nitrite at the time of NOB evolution and convert nitrate compounds (Dang & Chen, 2017).

7.1 Short-Term Nitrifying Activity

The short-term nitrifying activity test represented the dynamics of aquatic systems. This value may not properly represent in situ activity due to the addition of more substrate and aeration. It is critical to track activities over a period of time that does not allow the population to grow dramatically. Activity testing should last 12 h to avoid the latter. As a result, the test could be carried out over a long period of time, and

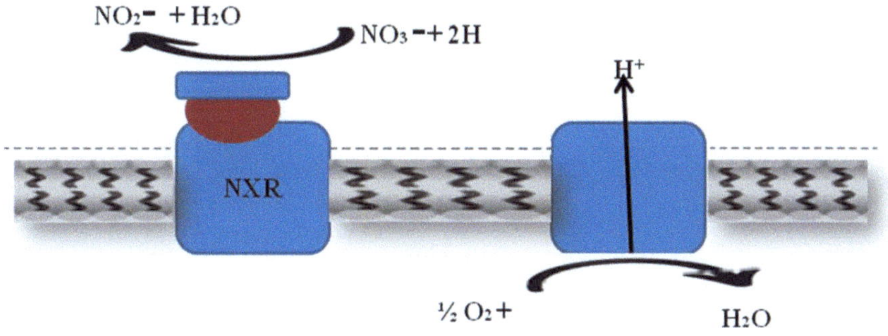

Fig. 2 The physiological mechanism of nitrite-oxidizing bacteria (Daims et al., 2016)

nitrite oxidation rates could be monitored at various nitrite concentrations. Perform four to five separate samplings every 1–2 h on a regular basis. The activity of NOB is reflected by a decrease in nitrite and a stoichiometric increase in nitrate under aerobic circumstances. The maximal activity per hour per g or mL of sample can be calculated by analyzing the slope of the drop in nitrite concentration over time (Manser et al., 2006).

7.2 Potential Nitrite-Oxidizing Activity

Natural collections are treated with substrate and incubated under favorable temperature and other circumstances in order to test their nitrifying potential. If mineralization happens at the same time, it is required to include an unaltered sample as a control. Prepare a duplicate and leave it to incubate for a few weeks. In contrast to short-term activity tests, potential activity can be assessed without agitation, and changes in nitrite and nitrate can be continuously tracked until the entire substrate has been oxidized. Anoxic microniches may favor denitrification when organic matter is present. To improve oxygen supply, the test vessels should be stirred. It is necessary to explore both the removal of nitrite and the production of nitrate. Denitrification might carry on indefinitely, consuming all labile organic matter (Spieck & Lipski, 2011).

7.3 Short-Term Activity of a Marine Biofilter Consisting of Colonized Plastic Biocarriers

5–10 colonized plastic biocarriers should be placed in 25–50 mL of sterile marine media with a 1–2 mM substrate concentration in separate test vessels and incubate on a shaker at an adequate temperature (17–28 °C) (170 rpm). Over the period of

8–12 h, or until the nitrite decline is clearly measurable, many samples should be collected (Spieck & Lipski, 2011).

7.4 Potential Nitrifying Activity of Permafrost Soil

In 100 mL of sterilized media with 0.75 mM nitrite substrate concentration in vessels, place 5 g of moist soil material. To make a uniform soil suspension, shake for 1 h at 120 rpm, then incubate at 17 °C without agitation. The time taken for continues growth might be as long as two weeks. It is preferable to determine the dry mass of a parallel soil sample in order to represent the activity/gram of soil (Wagner et al., 2002). Because the activity of liquid cultures is proportional to the number of cells present, population estimations are required. Although it has significant drawbacks, the most probable number technique was widely used in the past. When microcolonies are enumerated as single cells, this indirect numbering takes many weeks to months to complete and is filled with statistical errors. The recent developed molecular methods such as real-time fluorescence PCR enable fast identification of nitrifier populations in natural and anthropogenic contexts such as activated sludge or soils (Attard et al., 2010). NOB may be used to perform growth experiments to identify the optimal temperature, pH, and nitrite tolerance. Temperatures of 4–37 °C were used to establish the optimal temperature. It's vital to account for evaporation during long incubations at high temperatures. At 32 °C, nitrite oxidation was most efficient, taking just 7 days to complete, although activity was significantly reduced at 10 °C. At 4 and 37 °C, there was no action. It is possible to compare the slopes of the individual curves in order to get the best curve. Alternatively, when the original culture has completely oxidized all of the nitrite, the amount of substrate utilized (expressed as nitrite oxidation rate per day) can be compared (Kowalchuk & Stephen, 2001).

8 Chemotaxonomy Study of NOB

Chemotaxonomy is a useful tool for identifying and quantifying NOB in biofilms and pure cultures. The methods used for enriching, isolating, and characterizing NOB using various cultivation-based techniques are discussed in this chapter.

8.1 Biofilms

The use of micro-sensors in conjunction and molecular methods are quite beneficial for investigations into the microbial ecology of biofilms and specifically for the detection of nitrifying bacteria activity sites (Abeliovich, 2006). Schramm et al. (2000)

investigated the distribution of nitrifying bacteria from the genera *Nitrosomonas, Nitrosospira, Nitrobacter,* and *Nitrospina* in a membrane-bound biofilm system in which microsensors monitored oxygen, pH, nitrite, and nitrate gradients, and the nitrifying cultures along these gradients were identified and quantified using FISH and confocal laser scanning microscopy (Siripong & Rittmann, 2007). Ammonia oxidizers and representatives of the genus *Nitrobacter* dominated the oxic part of the biofilm, with *Nitrospira* sp. practically absent at the oxic-anoxic border. The cell populations of all nitrifiers were rather low in the completely anoxic portion of the biofilm. These findings point to an uncultured *Nitrospira spmicroaerophilic*'s activity as a factor impacting its environmental competitiveness (Berthe Corti & Fetzner, 2002).

8.2 Analytical Procedures and Oxidation Kinetics

Colorimetry can be used to determine nitrite and protein concentration (Han, 2018). Mass spectroscopy was used to track nitric oxide generation, nitrous oxide using gas chromatography, and oxygen consumption by using a Clark-electrode monitor (YSI model, Yellow Springs Instruments Co., USA). According to Sundermeyer-Klinger et al. (1984), membranes and nitrite oxidoreductase were extracted and were treated with 0.8% (w/v) n-octylglucoside for 12 h at 4 °C before being stacked on top of a linear 10–40% sucrose gradient to isolate cytochrome c oxidase. The gradients were centrifuged for 19 h at 300,000 × g. The cytochrome c oxidase was discovered to have a distinct green band. Francis and Becker's approach was used to identify heme proteins in polyacrylamide gels. In a spectrophotometer model MPS 2000 (Shimadzu) with a graphic printer PR-3, difference spectra were measured (Shimadzu) (Bock et al., 1990).

A micro-respiration system that included a one-channel oxygen sensor amplifier, a Clark-type oxygen microsensor, 2 ml glass chambers with glass stoppers, glass-coated magnets, and a stirring system can be used to measure nitrite-dependent oxygen consumption of cells from cultures. Cells were grown in mineral media as described by Nowka et al. (2015), cells were added to the chambers, which were then sealed and placed on a rack for stirring in a water bath at 37 °C. After a 15–30 min equilibration interval, the measurements were performed by syringing nitrite from stock solutions via a second capillary hole. The R-package "drc" was used to estimate nitrite oxidation kinetics from oxygen consumption rates at various defined nitrite concentrations, which fits Michaelis–Menten kinetics to the data with the equation: $V = (Vmax\ [S])/(Km + [S])$, where V represents activity, Vmax represents maximal specific activity (mol mg protein^{-1} h^{-1}), and Km represents the nitrite oxidation half-saturation constant (M), the nitrite concentration (M) is represented by [S]. The computed maximal activity was converted to mol $NO_2^{-\,cell-1}$ h^{-1} for cell-specific Vmax (Spieck et al., 2020).

8.3 Fame

Fatty acid profile analysis has proven to be a promising tool for swiftly and effectively distinguishing NOB. These methods can be used to track the differential enrichment of NOB populations during an enrichment procedure, evaluate the enrichment culture's increasing purity, and estimate the systematic orientation of the enhanced, close NOB population in the future. Specific lipid biomarkers can be used to differentiate the five known nitrite-oxidizing species (Alawi et al., 2007). *Nitrobacter* belonging to the Alphaproteobacteria, which means it, possesses cis-11-octadecenoic acid as its dominant fatty acid (18:1 cis-11). More than 70% of the fatty acid content of entire cells is made up of this lipid. There is also up to 20% cis-11,12-methyleneoctadecanoic acid (19:0 cyclo11-12) acid present. The total of the two fatty acids is normally between 80 and 90% of the time since this molecule is a derivative of 18:1 cis-11. The ratio of both lipids varies according to the growth phase and growth environment. Genera from Bradyrhizobiaceae, such as *Bradyrhizobium*, are closely linked to other Gram-positive bacteria, unlike most Gram-negative bacteria. Hydroxy fatty acids are absent in *Nitrobacter* species (Kruse et al., 2013). The lack of them is a worrisome sign for the NOB genus. *Nitrococcus* is a genus belongs to Gammaproteobacteria with a distinct profile. The 18:1 cis-11 ratio is much lower than in *Nitrobacter*, and hydroxy fatty acids are frequently detected, albeit the quantities aren't usually recognized. *Nitrotoga* is a recently discovered genus. A fatty acid composition that was clearly unique in one of a series of low-temperature produced enrichment cultures. In-depth examination into this unique community resulted in the identification of previously unknown phenomena. Finally, this nitrite-oxidizing bacteria has a new taxon. *Nitrospina* differs from other nitrite-oxidizing genera as it contains a considerable amount of tetradecanoic acid (14:0). The majority of *Nitrospira* strains include the hexadecenoic acid isomers cis-7 and cis-11, as well as 11-methyl-hexadecanoic acid (16:0 11-methyl), this species' branched chain lipid is the only one of its kind. In comparison, cis-9-hexadecenoic acid that is found in practically all bacteria as a constitutive membrane component is seldom seen in *Nitrospira* strains. Since these three biomarker lipids are present in different combinations and quantities in different *Nitrospira* species, this pattern can be exploited for chemotaxonomic inference at the species level (Lipski et al., 2001).

9 Contribution of NOB to Plant and Soil Health

The significance of microorganism-based on the management of soil and plant is widely debated in light of the expanding worldwide demand for both food and energy. In many ecosystems, nitrogen-fixing bacteria, in conjunction with plant and soil contributions, can provide the nitrogen demands of plants (Bender et al., 2016). Nitrogen-fixing bacteria, including symbiotic, associative, and free-living nitrogen-fixing microorganisms, can transform atmospheric nitrogen into ammonia, supplying nitrogen to the plant and increasing production. Furthermore, nitrogen fixation can promote microbial diversity and indirectly affect the soil nitrogen pool, both are beneficial to soil health. These nitrogen-fixing microbes use a range of nitrogen-fixing activities, all of which benefit the plant and soil health in the long term (De Araujo et al., 2008).

According to DNA studies, *Nitrospira* outweigh other NOB in acidic soils (Huang et al., 2021). *Nitrospira*, are rarely cultivated and their ecology is little understood. In addition, fresh study has discovered single bacteria inside *Nitrospira* that can oxidize both ammonia and nitrite, which was previously assumed to be a nitrite oxidizer. Comammox enrichment samples from terrestrial or acidic environments are few, despite their extensive distribution (Li et al., 2022).

For long-term enrichment, two separate continuous-feeding bioreactors capable of changing ammonia or nitrite content and pH were used. In Comammox, an excessive ammonium supply was discovered to be a crucial component (Sakoula et al., 2021). To avoid the buildup of free nitrous acid and cell growth inhibition at low pH, a modest dose of nitrite was also supplied. The amoA gene's amplicon sequencing revealed that two phylotypes belong to the Comammox clade (Huo et al., 2022). The enrichment sample also oxidizes ammonia at pH 4, which would be consistent with the very acidic tea field soil; however, this value is lesser than the active pH range of isolated acid-adapted nitrifiers (Kaur-Bhambra et al., 2022).

In agricultural contexts, nitrogen fertilizer is an important crop production management strategy that has a significant impact on the nitrifying population. In agricultural soils, AOB are more sensitive to nitrogen fertilization than AOA (Bello et al., 2021). Despite extensive research on the reaction of ammonia oxidizers to nitrogen fertilization, very less is known about the ecology of NOB in agricultural field. NOBs are classified into seven genera in the Proteobacteria, Nitrospirae, Nitrospinae, and Choloroflexi phylas (Hayatsu et al., 2021). In soils, Nitrobacter and Nitrospira are thought to be the two most important contributors. *Nitrobacter* spp. was found to be more susceptible to tillage practices, organic amendments (Shen et al., 2022).

10 Conclusion

NOBs are a challenging category of mostly or chemoautotrophs to grow. These bacteria provide a critical link between ammonia-oxidizing bacteria and various denitrifying microorganisms in the global nitrogen cycle. It eliminates nitrate using reduction mechanism and converts it into ammonia and complete the nitrogen cycle. Regardless of the difficulties in developing and nurturing most of these bacteria in the lab, NOB became the focus of considerable eco-physiological research over the years due to their critical role in global nitrogen cycling and acts as both primary contributors and scavengers of nitric oxides in the biosphere. Anaerobic ammonia oxidation, AOA (ammonia-oxidizing archea), and, most recently, total ammonia oxidation have all challenged our understanding of nitrification in recent decades

(Comammox). Before the discovery of *Nitrospira*, nitrification was thought to be a two-step process operated by chemolithoautotrophic AOB and NOB. According to its genome, Comammox (*Nitrospira*) has molecular machinery that can oxidize both ammonia and nitrite. Classical nitrite oxidizers and comammox are among the six sublineages of the phylum *Nitrospirae*. Single-step nitrification saves energy, and the presence of many metabolic pathways in *Nitrospira* is essential for the organism's establishment. The necessity of microorganism-based soil and plant health management has been widely debated in light of the growing worldwide demand for both food and energy. Nitrogen-fixing bacteria, in combination with plant and soil contributions, can provide the nitrogen needs of plants in various habitats.

References

Abeliovich, A. (2006). The nitrite-oxidizing bacteria. *The Prokaryotes*, *5*, 861–872.

Alawi, M., Lipski, A., Sanders, T., & Spieck, E. (2007). Cultivation of a novel cold-adapted nitrite oxidizing betaproteobacterium from the Siberian Arctic. *The ISME Journal, 1*(3), 256–264.

Albertsen, M., Saunders, A. M., Nielsen, K. L., & Nielsen, P. H. (2013). Metagenomes obtained by 'deep sequencing'–what do they tell about the enhanced biological phosphorus removal communities? *Water Science and Technology, 68*(9), 1959–1968.

Attard, E., Poly, F., Commeaux, C., Laurent, F., Terada, A., Smets, B. F., Recous, S., & Roux, X. L. (2010). Shifts between Nitrospira-and Nitrobacter-like nitrite oxidizers underlie the response of soil potential nitrite oxidation to changes in tillage practices. *Environmental Microbiology, 12*(2), 315–326.

Bello, M. O., Aigle, A., Meng, Y., Prosser, J. I., & Gubry-Rangin, C. (2021). Preferential temperature and ammonia concentration for in-situ growth of *Candidatus Nitrosocosmicus* ammonia oxidising archaea. *Soil Biology and Biochemistry*, *162*, 108405.

Bender, S. F., Wagg, C., & van der Heijden, M. G. (2016). An underground revolution: biodiversity and soil ecological engineering for agricultural sustainability. *Trends in Ecology and Evolution, 31*(6), 440–452.

Berthe-Corti, L., & Fetzner, S. (2002). Bacterial Metabolism of n-Alkanes and Ammonia under Oxic, Suboxic and Anoxic Conditions. *Acta Biotechnologica*, *22*(3–4), 299–336.

Bock, E. (1992). The genus Nitrobacter and related genera. *The Prokaryotes*, 2302–2309.

Bock, E., Sundermeyer-Klinger, H., & Stackebrandt, E. (1983). New facultative lithoautotrophic nitrite-oxidizing bacteria. *Archives of Microbiology, 136*, 281–284.

Bock, E., Wilderer, P. A., & Freitag, A. (1988). Growth of Nitrobacter in the absence of dissolved oxygen. *Water Research, 22*, 245–250.

Bock, E., Koops, H. P., Möller, U. C., & Rudert, M. (1990). A new facultatively nitrite oxidizing bacterium, Nitrobacter vulgaris sp. nov. *Archives of Microbiology*, *153*(2), 105–110.

Boon, B., & Laudelout, H. (1962). Kinetics of nitrite oxidation by *Nitrobacter winogradskyi. Biochemical Journal*, *85*(3), 440.

Daims, H., & Wagner, M. (2018). Nitrospira. *Trends in Microbiology, 26*(5), 462–463.

Daims, H., Lücker, S., & Wagner, M. (2016). A new perspective on microbes formerly known as nitrite-oxidizing bacteria. *Trends in Microbiology, 24*(9), 699–712.

Dang, H., & Chen, C. T. A. (2017). Ecological energetic perspectives on responses of nitrogen-transforming chemolithoautotrophic microbiota to changes in the marine environment. *Frontiers in Microbiology, 8*, 1246.

De Araujo, A. S. F., Figueiredo, M. D. V. B., & Monteiro, R. T. R. (2008). Potential of biological nitrogen fixation as indicator of soil pollution. *Nitrogen Fixation Research Progress*, 1–13.

Dworkin, M., & Gutnick, D. (2012). Sergei Winogradsky: A founder of modern microbiology and the first microbial ecologist. *FEMS Microbiology Reviews, 36*(2), 364–379.

Gruber, N. (2008) The marine nitrogen cycle: Overview and challenges. In D. G. Capone (Ed.), *Nitrogen in the marine environment* (2nd edn, pp. 1–50). Academic.

Han, S., Li, X., Luo, X., Wen, S., Chen, W., & Huang, Q. (2018). Nitrite-oxidizing bacteria community composition and diversity are influenced by fertilizer regimes, but are independent of the soil aggregate in acidic subtropical red soil. *Frontiers in Microbiology, 9*, 885.

Hayatsu, M., Katsuyama, C., & Tago, K. (2021). Overview of recent researches on nitrifying microorganisms in soil. *Soil Science and Plant Nutrition*, 1–14.

Huang, L., Chakrabarti, S., Cooper, J., Perez, A., John, S. M., Daroub, S. H., & Martens-Habbena, W. (2021). Ammonia-oxidizing archaea are integral to nitrogen cycling in a highly fertile agricultural soil. *ISME Communications*, *1*(1), 1–12.

Huo, D., Dang, Y., Sun, D., & Holmes, D. E. (2022). Efficient nitrogen removal from leachate by coupling Anammox and sulfur-siderite-driven denitrification. *Science of The Total Environment*, 154683.

Jacob, J., Nowka, B., Merten, V., Sanders, T., Spieck, E., & Dähnke, K. (2017). Oxidation kinetics and inverse isotope effect of marine nitrite-oxidizing isolates. *Aquatic Microbial Ecology, 80*(3), 289–300.

Janssen, P. H. (2006). Identifying the dominant soil bacterial taxa in libraries of 16S rRNA and 16S rRNA genes. *Applied and Environmental Microbiology, 72*(3), 1719–1728.

Kaur-Bhambra, J., Wardak, D. L., Prosser, J. I., & Gubry-Rangin, C. (2022). Revisiting plant biological nitrification inhibition efficiency using multiple archaeal and bacterial ammonia-oxidising cultures. *Biology and Fertility of Soils*, *58*(3), 241–249.

Könneke, M., Bernhard, A. E., José, R., Walker, C. B., Waterbury, J. B., & Stahl, D. A. (2005). Isolation of an autotrophic ammonia-oxidizing marine archaeon. *Nature, 437*(7058), 543–546.

Koper, T. E., El-Sheikh, A. F., Norton, J. M., & Klotz, M. G. (2004). Urease-encoding genes in ammonia-oxidizing bacteria. *Applied and Environmental Microbiology, 70*(4), 2342–2348.

Kowalchuk, G. A., & Stephen, J. R. (2001). Ammonia-oxidizing bacteria: A model for molecular microbial ecology. *Annual Reviews in Microbiology, 55*(1), 485–529.

Kruse, M., Zumbrägel, S., Bakker, E., Spieck, E., Eggers, T., & Lipski, A. (2013). The nitrite-oxidizing community in activated sludge from a municipal wastewater treatment plant determined by fatty acid methyl ester-stable isotope probing. *Systematic and Applied Microbiology, 36*(7), 517–524.

Lancaster, K. M., Caranto, J. D., Majer, S. H., & Smith, M. A. (2018). Alternative bioenergy: Updates to and challenges in nitrification metalloenzymology. *Joule, 2*(3), 421–441.

Larsen, P., Nielsen, J. L., Svendsen, T. C., & Nielsen, P. H. (2008). Adhesion characteristics of nitrifying bacteria in activated sludge. *Water Research, 42*(10–11), 2814–2826.

Lebedeva, E. V., Alawi, M., Fiencke, C., Namsaraev, B., Bock, E., & Spieck, E. (2005). Moderately thermophilic nitrifying bacteria from a hot spring of the Baikal rift zone. *FEMS Microbiology Ecology, 54*(2), 297–306.

Li, C., He, Z. Y., Hu, H. W., & He, J. Z. (2022). Niche specialization of comammox Nitrospira in terrestrial ecosystems: Oligotrophic or copiotrophic? *Critical Reviews in Environmental Science and Technology*, 1–16.

Lipski, A., Spieck, E., Makolla, A., & Altendorf, K. (2001). Fatty acid profiles of nitrite-oxidizing bacteria reflect theirphylogenetic heterogeneity. *Systematic and Applied Microbiology, 24*(3), 377–384.

Lücker, S., Wagner, M., Maixner, F., Pelletier, E., Koch, H., Vacherie, B., Rattei, T., Damsté, J. S. S., Spieck, E., Le Paslier, D., & Daims, H. (2010). A Nitrospira metagenome illuminates the physiology and evolution of globally important nitrite-oxidizing bacteria. *Proceedings of the National Academy of Sciences, 107*(30), 13479–13484.

Lücker, S., Schwarz, J., Gruber-Dorninger, C., Spieck, E., Wagner, M., & Daims, H. (2015). Nitrotoga-like bacteria are previously unrecognized key nitrite oxidizers in full-scale wastewater treatment plants. *The ISME Journal, 9*(3), 708–720.

Manser, R., Gujer, W., & Siegrist, H. (2006). Decay processes of nitrifying bacteria in biological wastewater treatment systems. *Water Research, 40*(12), 2416–2426.

Nocker, A., Burr, M., & Camper, A. K. (2007). Genotypic microbial community profiling: A critical technical review. *Microbial Ecology, 54*(2), 276–289.

Nowka, B., Off, S., Daims, H., & Spieck, E. (2015). Improved isolation strategies allowed the phenotypic differentiation of two Nitrospira strains from widespread phylogenetic lineages. *FEMS Microbiology Ecology, 91*(3), fiu031.

Ouyang, Y., & Norton, J. M. (2020). Nitrite oxidizer activity and community are more responsive than their abundance to ammonium-based fertilizer in an agricultural soil. *Frontiers in Microbiology*, 1736.

Pachiadaki, M. G., Sintes, E., Bergauer, K., Brown, J. M., Record, N. R., Swan, B. K., Mathyer, M. E., Hallam, S. J., Lopez-Garcia, P., Takaki, Y., & Nunoura, T. (2017). Major role of nitrite-oxidizing bacteria in dark ocean carbon fixation. *Science, 358*(6366), 1046–1051.

Park, S. J., Andrei, A. Ş, Bulzu, P. A., Kavagutti, V. S., Ghai, R., & Mosier, A. C. (2020). Expanded diversity and metabolic versatility of marine nitrite-oxidizing bacteria revealed by cultivation- and genomics-based approaches. *Applied and Environmental Microbiology, 86*(22), e01667-e1720.

Patel, P., Naik, A., & Sharma, A. (2022). Isolation and identification of nitrite-oxidizing microbes. In *Practical handbook on agricultural microbiology* (pp. 219–221). Humana.

Prosser, J. I. (1989). Autotrophic nitrification in bacteria. *Advances in Microbial Physiology, 30*, 125–181.

Rani, S., Koh, H. W., Rhee, S. K., Fujitani, H., & Park, S. J. (2017). Detection and diversity of the nitrite oxidoreductase alpha subunit (*nxrA*) gene of Nitrospina in marine sediments. *Microbial Ecology, 73*(1), 111–122.

Sakoula, D., Koch, H., Frank, J., Jetten, M. S., van Kessel, M. A., & Lücker, S. (2021). Enrichment and physiological characterization of a novel comammox Nitrospira indicates ammonium inhibition of complete nitrification. *The ISME Journal, 15*(4), 1010–1024.

Schmidt, E. L., & Belser, L. W. (1994). Autotrophic nitrifying bacteria. *Methods of Soil Analysis: Part 2 Microbiological and Biochemical Properties, 5*, 159–177.

Schramm, A., de Beer, D., Gieseke, A., & Amann, R. (2000). Microenvironments and distribution of nitrifying bacteria in a membrane-bound biofilm. *Environmental Microbiology, 2*, 680–686.

Shah, M. P. (2020). *Microbial bioremediation and biodegradation*. Springer.

Shah, M. P. (2021). *Removal of refractory pollutants from wastewater treatment plants*. CRC Press.

Shen, D., Qian, H., Liu, Y., Zhao, S., & Luo, X. (2022). Nitrifier community assembly and species co-existence in forest and meadow soils across four sites in a temperate to tropical region. *Applied Soil Ecology, 171*, 104342.

Siripong, S., & Rittmann, B. E. (2007). Diversity study of nitrifying bacteria in full-scale municipal wastewater treatment plants. *Water Research, 41*(5), 1110–1120.

Sliekers, A. O., Haaijer, S., Stafsnes, M. H., Kuenen, J. G., & Jetten, M. S. (2005). Competition and coexistence of aerobic ammonium-and nitrite-oxidizing bacteria at low oxygen concentrations. *Applied Microbiology and Biotechnology, 68*(6), 808–817.

Sorokin, D. Y., Lücker, S., Vejmelkova, D., Kostrikina, N. A., Kleerebezem, R., Rijpstra, W. I. C., Damsté, J. S. S., Le Paslier, D., Muyzer, G., Wagner, M., & van Loosdrecht, M. (2012). Nitrification expanded: discovery, physiology and genomics of a nitrite-oxidizing bacterium from the phylum Chloroflexi. *The ISME Journal, 6*(12), 2245–2256.

Spieck, E., & Bock, E. (2015). Nitrospina. *Bergey's manual of systematics of Archaea and bacteria* (pp. 1–7).

Spieck, E., & Lipski, A. (2011). Cultivation, growth physiology, and chemotaxonomy of nitrite-oxidizing bacteria. In *Methods in enzymology* (Vol. 486, pp. 109–130). Academic.

Spieck, E., Sass, K., Keuter, S., Hirschmann, S., Spohn, M., Indenbirken, D., Kop, L.F., Lücker, S., & Giaveno, A. (2020). Defining culture conditions for the hidden nitrite-oxidizing bacterium Nitrolancea. *Frontiers in Microbiology*, 1522.

Sundermeyer-Klinger, H., Meyer, W., Warninghoff, B., & Bock, E. (1984). Membrane-bound nitrite oxidoreductase of Nitrobacter: evidence for a nitrate reductase system. *Archives of Microbiology, 140*(2), 153–158.

Tao, Y., & Gao, D. W. (2012). Impact of ecological factors on anaerobic ammonia-oxidizing bacteria enrichments. *Environmental Engineering Science, 29*(6), 479–485.

Ushiki, N., Fujitani, H., Aoi, Y., & Tsuneda, S. (2013). Isolation of Nitrospira belonging to sublineage II from a wastewater treatment plant. *Microbes and Environments*, ME13042.

Vijayan, A., Vattiringal Jayadradhan, R. K., Pillai, D., Prasannan Geetha, P., Joseph, V., & Isaac Sarojini, B. S. (2021). Nitrospira as versatile nitrifiers: Taxonomy, ecophysiology, genome characteristics, growth, and metabolic diversity. *Journal of Basic Microbiology, 61*(2), 88–109.

Wagner, D., Spieck, E., Bock, E., & Pfeiffer, E. M. (2002). Microbial life in terrestrial permafrost: Methanogenesis and nitrification in gelisols as potentials for exobiological process. In *Astrobiology* (pp. 143–159). Springer.

Waheed, H., Hashmi, I., Naveed, A. K., & Khan, S. J. (2013). Molecular detection of microbial community in a nitrifying–denitrifying activated sludge system. *International Biodeterioration & Biodegradation, 85*, 527–532.

Wang, B., Zhao, J., Guo, Z., Ma, J., Xu, H., & Jia, Z. (2015). Differential contributions of ammonia oxidizers and nitrite oxidizers to nitrification in four paddy soils. *The ISME Journal, 9*(5), 1062–1075.

Watson, S. W., & Waterbury, J. B. (1971). Characteristics of two marine nitrite oxidizing bacteria, *Nitrospina gracilis* nov. gen. nov. sp. and *Nitrococcus mobilis* nov. gen. nov. sp. *Archiv für Mikrobiologie, 77*(3), 203–230.

Watson, S. W., Bock, E., Harms, H., Koops, H.-P.,Hooper, A. B. (1989). Nitrifying bacteria. In: R. G. E. Murray, D. J. Brenner, M. P Bryant, J. G. Holt, N. R. Krieg, J. W. Moulder, N. Pfennig, P. H. A. Sneath, J. T. Staley, & S. Williams (Eds.), *Bergey's Manual of Systematic Bacteriology* (Vol. 3, pp. 1808–1834). Baltimore, MD: Williams and Wilkins.

Woldendorp, J. W., & Laanbroek, H. J. (1989). Activity of nitrifiers in relation to nitrogen nutrition of plants in natural ecosystems. *Plant and Soil, 115*(2), 217–228.

Nitrogen Cycling in the Course of Biological Treatment of Wastewater in wetlands—An Analysis

Sonali Paul, Camellia Mazumder, Aditi Biswas, Aratrik Roy, and Susmita Mukherjee

Abstract Nitrogen is abundant in the atmosphere (almost 80%) and remains in highly stable and non-reactive forms, yet it is the sole source in the ecosystem. The reactive forms like ammonium, nitrite and nitrate act as nutrients and are, thus, very essential for plant growth, but their content is limited in the soil system. Hence, aerial nitrogen is utilized in the form of fertilizer for food production and supply as protein through the food web to various trophic levels. After the death of organisms, consortiums of microorganisms take the responsibility to return back the bound nitrogen to the atmosphere following some sequences of reactions. Hence, nitrogen cycling is important for maintaining nitrogen balance in the ecosystem. Nitrogen, especially the domestic wastewater is mostly found in organic form with negligible number of oxidative forms. The conventional wastewater treatment does not involve any separate nitrogen removal process. It is only removed by natural biological treatment process including steps of ammonification, nitrification, eutrophication and denitrification. Together with this, some bio-induced physical processes like sedimentation, volatilization, de ammonification (anammox) and sequestration might play a significant role in nitrogen balancing in the atmosphere. However, during the process of biological removal of nitrogen from wastewater it might release some amount of nitrous oxide gas, increasing the greenhouse gas footprint. Hence, it is urgently necessary to highlight the more sustainable pathways for nitrogen recovery and reusing in the course of conventional as well as natural wastewater treatment.

Keywords Molecular nitrogen · Nitrification · Denitrification · De ammonification · Aerobic · Anaerobic

---

S. Paul · C. Mazumder · A. Biswas · A. Roy · S. Mukherjee (✉)
University of Engineering & Management, University Area, Plot No. III, B/5, New Town Rd, Action Area III, Newtown, Kolkata, West Bengal 700160, India
e-mail: susmita.mukherjee@uem.edu.in

M. P. Shah (ed.), *Anammox Technology in Industrial Wastewater Treatment*,
https://doi.org/10.1007/978-981-99-3459-1_8

1 Introduction

Nitrogen is important because it is a major part for the biochemical reactions in the living body. From enzymatic to hormonal changes, every procedure involves nitrogen and its molecules in different chemical forms. So, it plays a major role in the biochemistry of living organisms. Abiotic components also play a major part in nitrogen cycling, its production and its utilization. Different sources of nitrogen production from biotic and abiotic components contribute to adding nitrogen to the environment thus adding the essential number of compounds of nitrogen to the atmosphere. The nitrogen cycling is a biogeochemical process that shows the movement of nitrogen from the atmosphere by living matter through the biosphere and again into the atmosphere in which the nitrogen is converted into its different forms, involving land, water and air ecosystems. Not just ecosystem but also in human body—the main central dogma in any living body consisting of the major macromolecules like proteins and nucleic acids are mostly aromatic chains of carbon, nitrogen and oxygen. They have an abundance of nitrogen molecules and compounds that play vital roles in the system's metabolism; proteins are the main components of all these procedures. Proteins are the most abundant molecules in a living cell, amino acids being its building block and the amino group as a vital portion for its general structure. This amino group consisting of three hydrogen molecules and one nitrogen molecule forms the general amino acid structure along with the carboxyl group. Amino acids are ions also known as zwitterions in aqueous solution and behave as both acids and bases as they are amphoteric. Proteins are formed by polymers of amino acids which are formed by intermolecular peptide bonds. These peptide bonds are formed when the α-carboxyl group of an amino acid is linked to the α-amino group of another amino acid through an amide linkage called peptide bond, thus releasing a water molecule. These polypeptides containing nitrogen groups are the basis of all the hormonal and enzymatic activities in the biochemical process of a living organism.

Wastewater is a major source for contributing ammonia and nitrogen by-products; thus, they contribute largely to the nitrogen content in the ecosystem and atmosphere in both gaseous and aqueous form. Wastewater includes all the liquid waste produced after using clean water supply for a diversifying co-existence. Most wastewater holds large amounts of readily biodegradable molecules. A wastewater treatment plant is a facility wherein industrial wastewater is handled with the elimination of pollutants. Nitrogen compounds being the paramount constituents of wastewater present in various concentrations of water sources. Now, many wastewater treatment plants around the globe discharge the wastewater into the surrounding without eliminating the reactive nitrogen from it. The discharge of disproportionate reactive nitrogen not only brings ruinous effects on human health but also can put up pollution in air and water bodies and create destruction in the complex ecosystem. To steer clear of adverse effects of surplus reactive nitrogen in the human habitat, tertiary wastewater treatment measures that ensure the deletion of reactive nitrogen varieties need to be carried out.

2 Importance of Nitrogen Cycling in Ecosystem

Nitrogen is vital because it is a key component of metabolic reactions in living organisms. Every procedure requires nitrogen and its molecules in various chemical forms, from enzymatic to hormonal modifications. As a result, it has a significant impact on the biochemistry of living creatures. Abiotic elements have an important role in nitrogen cycle, production and usage. Different sources of nitrogen generation from biotic and abiotic components contribute to nitrogen addition to the environment, hence supplying the necessary quantity of nitrogen compounds to the atmosphere. Nitrogen cycling is a biogeochemical process that involves land, water and air ecosystems and depicts the transfer of nitrogen from the atmosphere through the biosphere and back into the atmosphere, where it is changed into various forms (Lie et al., 2021). The nitrogen cycle's amazing modifications have led to a wide range of changes (Galloway et al., 2003).

Nitrogen in atmosphere: Nitrogen takes up more mass than any other gas that includes 21% oxygen and 1% of neon, carbon dioxide, krypton, methane, hydrogen, helium, argon in the atmosphere. Earth's atmosphere is majorly made up of five layers: the exosphere, thermosphere, mesosphere, stratosphere and the troposphere. The troposphere is the widest layer in the atmosphere and the closest to the earth's surface. Nitrogen (approximately 78%) is present in the troposphere as it is the closest layer to the surface (about 10–18 km above the earth's surface) and also the densest so nitrogen that is mostly entrapped in the rubble forming the earth, cannot escape to the other layers of the atmosphere because of its density, as studies say (Lie et al., 2021).

Nitrogen cycling in Terrestrials: The nitrogen present in the tropospheric region of the atmosphere is entering into the terrestrial region through biological or industrial processes or through rain. After entering it is going through processes of conversion by nitrogen fixation, nitrification and finally getting out again into the atmosphere by the process of denitrification. Precipitation is the starting point of nitrogen in the ground soil (rain, lightning). Nitrogen in the air isn't the only source of nitrates. They may also be produced by nitrifying bacteria in the soil converting nitrate, which is often used in fertilizers. Nitrogen in the soil can also be converted to nitrates by certain root nodules. Plants use nitrates taken from the soil to build up protein (Adams, 2003). When animals eat these plants, such as cows, they utilize it to make animal protein. When animals (cows) defecate, pee, or die, the nitrogen in their urea, excreta, or carcasses is broken down by decomposers and reintroduced into the soil as ammonia (Gwak et al., 2020) which then gets converted into nitrates by nitrifying bacteria and then into again nitrogen gas by denitrifying bacteria as mentioned in the Fig. 1. Livestock, fertilizer, soils, wildfires and slash burning, industries, transportation, the oceans, humans, pets, wild animals, plus waste disposal and recycling operations are all common sources of ammonia (Anderson et al., 2003).

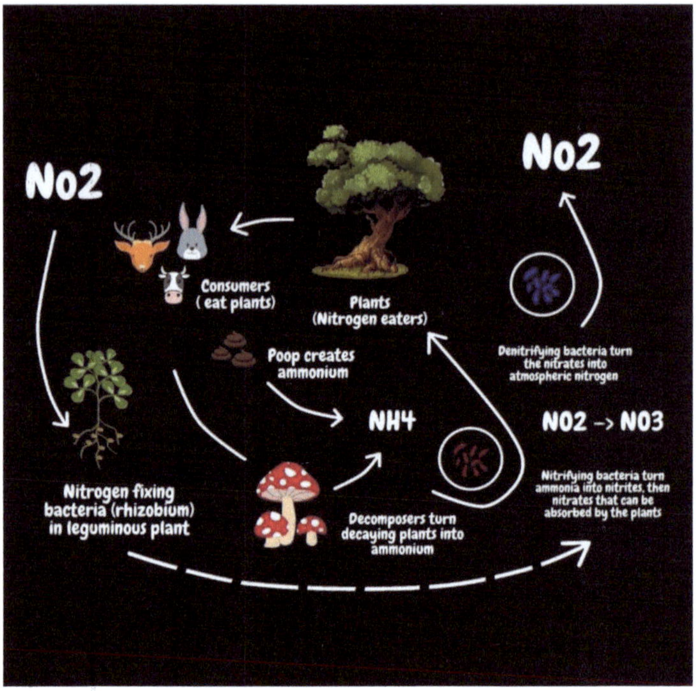

Fig. 1 Nitrogen cycling in ecosystem

3 Importance of Nitrogen Cycling for Living Organisms

Nitrogen is a keystone of amino acids, which are the constituents of proteins, and nucleic acids, such as DNA, which disseminates genetic details to sequential generations of organisms, nitrogen is requisite for all living entities and such studies are showing the results in herbivores too (Kampman et al., 2014). Proteins are the most numerous molecules in a live cell, and amino acids, which serve as its building blocks, play an important role in its overall structure so its presence is important in our diet which is calculated by a protein-to-nitrogen factor as devised in the recent studies (Mariotti & Mirand, 2008), this is found to be 5 in seaweeds (Angell et al., 2016). Along with the carboxyl group, the overall amino acid structure is formed by this amino group, which consists of three hydrogen molecules and one nitrogen molecule. In aqueous solution, amino acids are ions known as zwitter ions that act as both acids and bases due to their amphoteric nature. Proteins are polymers of amino acids linked together by intermolecular peptide bonds. These peptide bonds are generated when an amino acids' α-carboxyl group is attached to the α-amino group of another amino acid via an amide linkage known as a peptide bond, releasing a water molecule. The growth and the metabolism of the living body are dependent on these units called nucleic acids and the process is often considered as the central dogma with proteins as its final product. The nucleic acids or the DNA and RNA molecules are heterocyclic aromatic rings. Aromatic rings are generally carbon and nitrogen-containing organic rings. These bases form nucleosides with a pentose sugar. The nucleoside with one or more phosphate groups forms the nucleotide. Nitrogen is used for the creation of amino acids, then they are synthesized into proteins in human bodies. Proteins play a major role in athletes as understood by some scientific research (Hoffman & Falvo, 2004).

Nitrogen is an integral component of amino acids, which further acts as building blocks of proteins. Proteins are necessary for the proper functioning of animal bodies. Nitrogen influences very important bodily functions, such as immune response, growth, hormonal action, etc. Severe deficiency of proteins leads to threatening diseases such as kwashiorkor and marasmus (Watford & Wu, 2018) Apart from that, they form a significant portion of nucleic acids, which constitutes genetic material. As known by all, without genetic material, there would be no life. If the human body is considered, it's totally evident that nitrogen plays a crucial role in growth and development. Nitrogen is needed for tissue repair, cell replacement and for the formation of new healthy cells. It has been observed that nitrogen deficiency leads to a strong negative effect on the human body, often leading to serious malnutrition, as few research work suggests (Hoghooghi et al., 2021).

The nitrogen cycle is indispensable for all living entities. Bacteria employ nitrogen from the atmosphere to manufacture nutrients in the soil. Plants need these manufactured nutrients to develop properly. Animals and humans utilize nitrogen built by plants. It is extant in a miscellaneous condition in the human body, unlike the nitrogen present in the atmosphere. Proteins, nucleic acids and other organic molecules all encompass nitrogen and thus is conferred from recent studies (Denancé et al., 2014). Plants or mostly the leguminous plants are only capable organisms for nitrogen fixation in the soil which further cater to other animals and living beings. From these leguminous plants, the living organisms receive the nitrogen content necessary for their metabolism and growth and regeneration activities. Thus, the whole food chain starts from plants providing nitrogen which is passed on to the herbivores, which provide nutrients to the secondary consumers and finally the tertiary consumers leading to scavengers and the microbes that decompose these dead wastes to the soil. These leguminous plants have special procedures to fulfill the requirements. Nitrogen fixation occurs in legumes in specialized entities known as root nodules. The nodules form as a result of the interaction of the bacteria *Rhizobium* with the legume roots. The metabolic processes involved in nitrogen fixing are the same. However, bean nodules contain a unique protein known as LEGHEMOGLOBIN. Leghemoglobin production is the outcome of symbiosis because neither bacteria nor bean plants can produce the protein on their own. It was recently discovered that a variety of host genes are involved in this process. In addition to leghemoglobin, a collection of proteins known as nodulins are generated, which has a great impact on leguminous plants, as the paper says (Kraft et al., 2014).

Nitrogen Fixation: It is a chemical process through which the nitrogen absorbed by the plants is converted into the form, i.e., ammonia (Hoffman et al., 2014), which is usable by the plants and other biological organisms like microorganisms/bacteria, for the production of amino acid, protein for their survival. In this process, the atmospheric nitrogen present in the plants gets subsequently transformed to ammonia, then into nitrites and then into nitrates through symbiotic bacteria (*Rhizobium*) found in leguminous plant root nodules as it is needed by the plants and the plants provide nutrition for the bacteria. These bacteria have the nitrogenase enzyme that has the capability of combining nitrogen gas with hydrogen to create ammonia, as discussed in this paper (Chen et al., 2017).

Nitrification: Nitrification is a biological process where the ammonia is oxidized into nitrites, these nitrites are again oxidized into nitrates on the other hand direct oxidation of ammonia into nitrate also defines the process of nitrification (Liu et al., 2014). This process is carried out in soil by autotrophic nitrifying bacteria (Fink et al., 2011) like *Pseudomonas spp.* under aerobic conditions and recently it was observed that bacteria are capable of complete nitrification of ammonia to nitrate (Lei et al., 2019). The nitrogen provided to the plants predominantly as nitrate or the nitrogen provided to the plants as ammonium salts as fertilizer is all converted to nitrates through the process of nitrification, where the nitrifying bacteria transforms those ammonia into nitrites and subsequently into nitrates. Nitrification can be performed chemically too which also acts as a routine step in Biological Oxygen Demand count (Liu et al., 2014). This final product of nitrification is used as a vital source of nitrogen by the plant body, as conferred by research (Liu et al., 2014).

Denitrification: This is the final microbially mediated process of nitrogen cycle, where the nitrates produced after the process of nitrification are used as a form of energy by anaerobic denitrifying bacteria like *Pseudomonas aeruginosa*, and gets converted back into nitrogen gas which reenters the atmosphere. This process takes place in the absence of oxygen, i.e., places like deep in the soil or muddy environments. In some ecosystems, denitrification plays a very big role to decrease the rate of nitrates inside the soil as that nitrate can get in vegetables which when consumed by human being can cause the risk of cancer, as understood from this paper (Rajta et al., 2020). Denitrification is a universal process taking part in nitrogen cycling of both terrestrial as well as aquatic ecosystems keeping the percentage of nitrogen in atmosphere stable and utilizing some engineering techniques too (Lei et al., 2014).

4 Nitrogen Cycling in a Water Body

The nitrogen cycle in land/soil is identical to the nitrogen cycle in any water body or marine habitat. Water absorbs nitrogen from the atmosphere, and nitrogen-containing molecules settle as rocks on the ocean floor. Many animals are unable to dissolve the strong nitrogen bond. However, only a few bacteria are capable of oxidizing the nitrogen molecule and converting it to ammonia (Beeckman et al., 2018). Ammonia may be absorbed by phytoplankton plants. Some bacteria ingest ammonia and produce nitrites as a result. The silt settles on the ocean floor whenever fishes or marine animals die. The bacteria disintegrate them, releasing ammonia, which is then transformed to nitrates via nitrification, and the cycle continues to repeat. The nitrates are subsequently transformed into nitrates, which are then utilized by another marine microbe. Nitrification is the procedure of transmogrifying ammonia into nitrates. Phytoplankton is consumed by larger species such as whales, fish and other animals. Apart from its importance in biological organisms, the nitrogen cycle has an important role in various water bodies. It's necessary to learn about this since multiple environmental phenomena are being affected by the actions of humans, which indirectly affects nitrogen in the water (Landner, 2013).

5 Wastewater as a Rich Source of Nitrogen

Wastewater is a polluted form of water that comes from day-to-day activities. This includes wastewater from domestic and residential sources as well as non-domestic sources (Roy & Kent, 2007). Depending on its origin, the volume and characteristics of wastewater vary to a great extent as shown in Table 1. The best part about wastewater is that it can be utilized to extract various valuable resources, with divergent qualities that can be suited to a wide array of industries worldwide. Now wastewater endangers public health plus industrial processes explicitly or implicitly. As a result, a speedy, easy, environmentally safe, effective and efficient technology for removing contaminants from municipal and industrial wastewater effluent is required (Ward, 2003). Nitrogen cycle is a very vital chemical process in terms of the biogeochemistry of the environment. Large amount of natural nitrogen flows from the atmosphere to the tellurian and marine biosphere, where it is acted upon by various microorganisms and physicochemical factors, which further results into the formation of various compounds and amino acids. Nitrogen has a major role to play in the biosynthesis of vital cellular components such as proteins and nucleic acids. While atmospheric nitrogen serves as the common source for multiple bacterial species (e.g. nitrogen-fixing bacteria and archaea), other organisms depend on a complex form like nitrates, etc.

5.1 By-Products of Nitrogen Retrieval from Wastewater

Nitrogen retrieval from wastewater is undoubtedly a daunting process. There are several by-products that are released at the end of the recovery process. Sludge, grit, biogas (methane, carbon dioxide, etc.), hydrogen sulfides are some of the most common by-products found at the end of the process. Apart from nitrogen, the other byproducts also have some sort of utility. For instance, methane is used in the automobile and fuel industry, fertilizer industry and other similarly vital industries. Carbon dioxide is being widely accepted as a means of pH control in water treatment plants. This is because it helps in maintaining the alkalinity and prevents formation of harmful products at the end of the process as concluded from the study (Rao et al., 2007).

Table 1 Wastewater discharge from water bodies

Components of Wastewater	
Potentially Hazardous	11%
Organics	23%
Paper	15%
Plastics	9%
Ferrous metals	5%
Non - ferrous metals	1%
Glass	2%
Naples and sanitary	3%
Textiles	4%
Rubble, concrete, etc	12%
Rubber	1%
Timber	14%

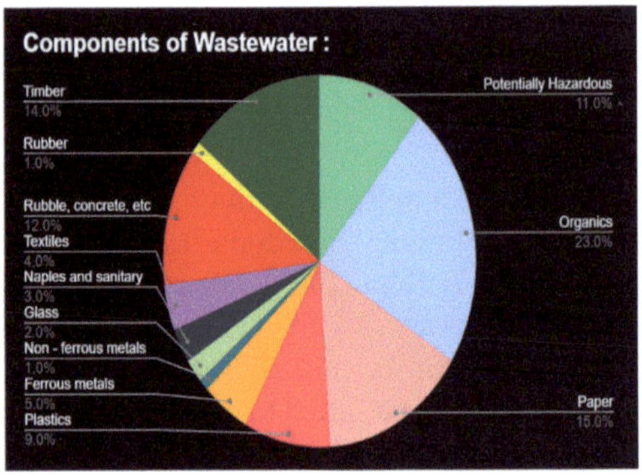

Aeration is a central process that involves a certain microbe that acts on the sample, which undergoes nitrogen fixation and thus produces nitrogen in gaseous form to release into the atmosphere and complete the cycle of nitrogen. Aeration can be defined as the process by which air is added to the sewage to facilitate the bio-decomposition of the organic components present. As discussed before, bacteria (specially the aerobic ones) have a major role in treatment of wastewater. Even though most wastewater treatment was mostly done in traditional anaerobic–aerobic treatment facilities, high-rate anaerobic–aerobic bioreactors have become more popular in recent years for wastewaters having high chemical oxygen demand (COD) (Sliekers et al., 2002). Such bacterial species utilize the free oxygen to break down the pollutants and release energy. This energy is further used for multiplication by the bacteria. In the absence of oxygen, the biodegradation of organic matter is very slow and inefficient. Studies have revealed that aerobic treatment is most useful for low-strength wastewaters. Some examples of obligate aerobes include *Mycobacterium, Bacillus, Pseudomonas, Nocardia* etc. In case of nitrogen fixation, which is a vital portion of wastewater treatment, such aerobic bacteria come to aid (e.g., *Azospirillum*). Other examples of aerobic nitrogen-fixing bacteria include *Rhizobium* and *Azotobacter*. They are involved with the fixation of nitrogen in plant and cereal families.

5.2 Nitrogen Discharges from Forests and Industries

Nitrogen (N) is a limiting nutrient for forest development, a driver of terrestrial ecosystems and a driver of eutrophication within marine ecosystems in the boreal environment. Previously examined studies (Roy et al., 2018) on forest nitrogen patterns in northern settings and discussed how management and atmospheric change affect internal cycling and export. Resolving the nutritional relevance of various nitrogen types to plants and determining how tree–mycorrhizal connections affect nitrogen limitations are two important research topics (Roy et al., 2018). Furthermore, determining how above and belowground ecological compartments affect forest reactions to nitrogen inputs from external sources remains a significant issue. Eventually, in maintained forest landscapes, forestry creates a mosaic of successional regions with varying degrees of nitrogen input, ecological demand and hydrological loss. The interplay between these mechanisms has an impact on the temporal patterns of river water composition as well as forest growth's long-term survival. Finally, regulating forests to meet rising biomass production requirements while limiting environmental degradation would necessitate multi-scale and multidisciplinary approaches to landscape nitrogen dynamics (Roy et al., 2018).

A deeper examination of real nitrogen releases from the forest sector, therefore, appears to be critical. The major focus of attention in waste effluents from pulp and paper mills has been on the presence of organic compounds in solution or as suspended debris for many years. The presence of plant nutrients in pulp and paper mill wastes has only lately been focused since significant reductions in the discharges of oxygen-consuming organic matter have been achieved. Among the plant nutrients found in these effluents, phosphorus has been the topic of more in-depth research previously done (Washington State Department of Health, 2014).

Some information on phosphorus releases from the forest sector are available. In the case of nitrogen, the estimations available in the literature are fairly inadequate. Organic nitrogen makes up the majority of the nitrogen in papermaking mill effluents, which is not immediately available to the primary producers in the receiving body of water. In comparison to this, sulfite mills, which use ammonia as a base in the cooking fluid, are the polar opposite. Significant ammonium nitrogen emissions from such mills are possible (Poh et al., 2016).

5.3 Nitrogen Discharge from Fertilizer Industry

We deciphered from the paper (Roy et al., 2018) that fertilizers provide the nutrients to plants that they require for survival and prosperity. Plants live, develop and reproduce by absorbing water from soil and nutrients All plants obtain their additional required minerals from the place where they grow—which is primarily soil, though 90–95% of the plants' dry matter comes from oxygen (O2), hydrogen (H2), Carbon (C). Plants need oxygen that make up 90–95% of all plants' dry matter, other components such as nitrogen, phosphorus, potassium, calcium, magnesium and sulfur. The remaining components are classified as major primary nutrients and secondary nutrients like (Roy et al., 2018) and micro-nutrients (boron, chlorine, copper, iron, manganese, molybdenum and zinc) in much lesser proportions. Organics, alcohols, ammonia, nitrates, phosphorus, heavy metals including cadmium, and suspended particles are all found in chemical fertilizer wastewater (Roy & Kent, 2007). As seen from the article-to avoid land degradation and preserve the fertility and production of the soil, the optimal nutrients required by the crops should be replenished to them through mineral fertilizers also, the use of fertilizer results in the increased agricultural outputs contributing to physicochemical characteristics like soil organic matter maintenance, moisture holding capacity, biological nitrogen fixation, soil depletion regulation and less extensive land use, among other benefits to producers, customers and the environment (Roy et al., 2018).

The fertilizer industry is complex, with a broad variety of fertilizer types and grades to choose from. The majority of the aqueous waste generated by the fertilizer industry, on the other hand, comes from the manufacture of only a few important products. Because water recycling, small process differences, and operational philosophies may result in a large range of waste stream compositions and volumes for a given product amongst different fertilizer plants, defining the quantities and compositions of the various waste streams is extremely challenging. Furthermore, process spills account for a significant portion of overall waste discharges due to their unpredictable frequency and quantity. Process water makes up approximately 20–25% of the water utilized by the fertilizer industry on average (Watford & Wu, 2018), but it comprises the majority of the contaminants produced by different processes like pharmaceuticals (Tang et al., 2011).

Various industries such as the sugar industry, starch industry and vegetable oil industry are responsible for discharging a great quantity of wastewater. The wastewater from the sugar industry has a high number of contaminants, which includes mud, bagasse, sludge, etc. This is why different advanced technologies are being used to treat wastewater from the sugar industry, as suggested by previous works (Sponseller et al., 2016). Most necessary components of wastewater are retrieved from high organic loads of sugar. Wastes are obtained from power plants, distillation of spirit from molasses and crushing of the cane. The starch industry has an elevated level of organic load and easily degradable components. Starch water has a high content of nitrogen along with phosphorus. Numerous types of bacteria convert atmospheric nitrogen (N2) into useful forms such as NO_{2-} and NO_{3-}. The category of denitrifying bacteria (Kampman et al., 2014) is responsible for turning NO_{2-} back into atmospheric nitrogen. This establishes a clear relationship between nitrogen, its forms and the ecosystem. Anaerobic bacteria are heavily involved in wastewater treatment, since they act on the sludge and reduce their volume, producing methane gas along the process. They are the best option for treating high-strength wastewater using mechanistic and meta-heuristic methods as we could confer from the research works (Puijenbroek et al., 2019).

6 Septic Systems Might Cause Pollutant Productions like Nitrogen in Wastewater

Household sources of wastewater are plenty (Gao et al., 2014). They mainly include sink water, water from laundry, kitchen and shower, toilets, etc. There are many elements in household discharges that serve as sources of nitrogen. Domestic sewages have several nitrogenous compounds. Organic matter such as vegetable and fruit peels, excreta, eggshells, and other wastes generate high amounts of nitrogen. Apart from that, the septic systems are a common feature in many localities, used to treat mainly human wastes (Bardgett et al., 1998). Compared to the effluents from industry, domestic wastewater has a smaller number of harmful pollutants, thus minimizing negative effects on the environment. Wastewater treatment facilities handle water from homes and businesses, which contains nitrogen and phosphorus from human waste, food, and some soaps and detergents (Annavajhala et al., 2018). If not properly managed, septic systems can rapidly become a source of nutrient contamination. Wastewater management facilities can very quickly turn into an origin of nutrient pollution if proper measures are not taken to maintain it. They supply elements such as nitrogen, phosphorus from hum food and wastes, detergents and soaps from household and industrial runoffs, etc. (United States Environmental Protection, 2021).

When the scum and particles are pumped out of the septic tank, a tiny quantity of nitrogen that enters the tank is eliminated. In most onsite sewage structures, oxygen-loving bacteria convert ammonium to nitrate within the drain discipline. This procedure is referred to as "nitrification" the effluent becomes "nitrified" superior structures that aerate and recirculate wastewater can dispose of even greater nitrogen (as much as 60%) (Washington State Department of Health, 2014). Increased nitrogen and phosphorus concentrations can be discharged into nearby water bodies as well as groundwater when a septic system is inadequately maintained. Septic systems are projected to fail 10–20% of the time over their service lives. Septic system failure is commonly caused by aged infrastructure, poor planning, saturation with far too much effluent in too little time, and inadequate maintenance. In most situations, burghers are amenable to ensuring their septic systems. Burghers should safeguard and maintain their system by having their system examined on a regular basis and having their tank pumped as needed, making good use of water, household hazardous garbage should not be disposed of in sinks or toilets and avoiding automobiles or putting large objects on their drain fields, among other things, these were concluded from the papers and past studies (United States Environmental Protection, 2021).

7 Biological Treatment of Wastewater in Wetlands and Nitrogen Cycling

Wetlands are the most diverse ecological landscapes and an important ecosystem for all the living and microorganisms. The wetland ecosystem plays a critical role in the world contributing to the whole biogeochemical cycle and as the natural converter of pollutants from nitrogen sources like wastewater. One of its important applications is to facilitate better wastewater recovery and recycling (Lee et al., 2009; Zhou et al., 2010a). In fact, they are often considered as one of the most productive ecosystems, exceeding the impact of terrestrial and aquatic ecosystems. Due to their diverse landscape, they contribute towards maintaining the balance of nature. Now, these ecosystems also have a vital function of acting as the source or sink for nitrogen, which has a wide range of applications in the world of environmental science. Nitrogen affects the productivity and functioning of wetland ecosystems. Thereby, it becomes all the more necessary to pay attention to the cycling of nitrogen in wetlands (Zhou et al., 2010b).

In wetlands, ecosystem nitrogen fixation takes place by both the biotic and abiotic components where the inorganic components are collected by the microbes and transformed into organic substances, which are absorbed by the soil or the wetlands making it a rich source of organic nitrogen. Now, the cycling and release of nitrogen in wetlands involve some crucial processes as observed in recent papers (DeBusk, 2003), which are interconnected among themselves including terrestrial run-offs, biological nitrogen fixation often referred as biofixation and wet deposition in soil due to rain or snow. Firstly, owing to the concentration gradient, diffusion of surface water takes place in wetlands, which allows seeping of the water deep below into the soil. Now, this water usually contains the inorganic forms of nitrogen, like ammonium and nitrates, which are absorbed and taken up by the wetland plants. The organic compost formed by the action of microorganisms on fallen leaves, stems and other plant parts, acts as a source for nitrogenous compounds. This is the time when the microbes participate in nitrification and denitrification (Yin et al., 2015) to form nitrogen, which finally gets stored in the soil. A better rate of nitrogen retention enhances the fertility of the soil, thus avoiding the over usage of harmful fertilizers (Roy & Kent, 2007). Also, the nitrogen in wetlands is stored for long due to peat accretion which contains the organic nitrogen in sources like plants and microbes. Thus, it is incurred that when the water moves to the downstream ecosystem it becomes a rich source of nitrogen replenishing the nitrogen deficiency in the fertile soil and in the land wherever nitrogen components were necessary (Wolf & Ahn, 2011).

The importance of wetlands in nitrogen cycling is because wetlands are a rich source of biodiversity, which is very useful in natural biogeochemical cycles. Also due to nitrogen deposit, it acts as a very good fertilizer (Gwak et al., 2020). Wastewater is a rich source of nitrogen but the environment in the wastewater is anaerobic, hence nitrogen is mostly in NH_4^+ form, which is the initial step in the nitrogen degradation pathway. The ammonium ion needs to be oxidized for converting it into the usable form, hence nitrogen cycling demands oxygen as an important input in the degradation pathway (Das Gupta et al., 2016). The ammonium form of nitrogen can get converted into its organic form and get sedimented at the bottom soil of the wetland or also get stored by the microorganisms (Shapley, 2009). On the contrary, the ammonium ion can get further oxidized into nitrate to be absorbed as nitrate salt by the plants. Thus, in wetland ecosystems, nitrogen can get fixed by both ecologically or biologically as nitrate salts in the soil (Zhou et al., 2012).

Throughout this, we could realize that the procedure of nitrogen cycling follows the steps of diffusion, decomposition and chemical procedures like denitrification and nitrification, which have been used for treating waste water biologically in more sustainable methods (Wong & GeoffBarfor, 2003). Wastewater being a rich source of nitrogen and its ammonium components contributes largely to the source of nitrogen cycling in wetlands because wetland soil mainly exists in the form of organic nitrogen also the wastewater can be treated by landfill leachate (Wiesmann & Wiesmann, 1994). This organic nitrogen comes from the plant uptake, the dead and decayed waste and in case of biological wastewater treatment, the source is wastewater, containing majorly nitrogen. Thus, proving that using nitrogen cycling, which takes place in wetland soil, is an effective biological treatment of wastewater with additional benefits of lesser waste production and economic solutions as shown in a combined isotope and acetylene approach as evident from the recent sources (Bateman & Baggs, 2005; Zhou et al., 2010c).

8 Conclusion

Throughout the article, this idea was realized that nitrogen is an important element for all the biological, biogeochemical processes and all the chemical procedures involving both biotic and abiotic environments. Being such an important element, it becomes the demand of the day to retrieve it from every viable source, may it be from industrial wastewater or domestic discharge. The role of various microorganisms must not be ignored in the entire nitrogen cycling process; the organic components can be further broken into simpler elements, credits to them. Nitrogen takes different forms in different media like nitrogen dioxide in air, ammonia/ammonium ion, nitrates and nitrites in water and soil nitrogenous bases in the macromolecules or the nucleic acids. Nitrogen cycling is a process of replenishing the environmental nitrogen pool. Lots of research are being done on areas involving nitrogen cycling and more viable pathways are being studied for the recovery of nitrogen from wastewater and sewage, without causing any damage to the environment. The use of microorganisms has been quite beneficial in terms of the quality of nitrogen recovered and the quantity. The wastewater sector is now in a period of transformation. It also opens up possibilities for rethinking nitrogen management in wastewater treatment systems by including innovative nitrogen removal processes like de-ammonification or shorter nitrification–denitrification. The present work through extensive discussion reveals different aspects of nitrogen cycling and its utilization in the ecosystem.

References

Adams, M. B. (2003). Ecological issues related to N deposition to natural ecosystems: Research needs. *Environment International, 29*, 89–199.

Anderson, N., Strader, R., & Davidson, C. (2003). Airborne reduced nitrogen: Ammonia emissions from agriculture and other sources. *Environment International, 29*, 277–286.

Angell A. R., Mata L., de Nys, R. et al (2016). The protein content of seaweeds: a universal nitrogen-to-protein conversion factor of five. *Journal of Applied Phycology*, (28), 511–524.

Annavajhala, M. K., Kapoor, V., Santo-Domingo, J., & Chandran, K. (2018). Comammox functionality identified in diverse engineered biological wastewater treatment systems. *Environmental Science and Technology Letters, 5*(2), 110–116.

Bardgett, R., Wardle, D. A., & Yeates, G. (1998). Linking above-ground and below-ground interactions: How plant responses to foliar herbivory influence soil organisms. *Soil Biology and Biochemistry, 30*, 1867–1878.

Bateman, E. J., & Baggs, E. M. (2005). Contributions of nitrification and denitrification to N_2O emissions from soils at different water-filled pore space. *Biology and Fertility of Soils, 41*, 379–388.

Beeckman, F., Motte, H., & Beeckman, T. (2018). Nitrification in agricultural soils: Impact, actors and mitigation. *Current Opinion in Biotechnology, 50*, 166–173.

Chen, H., Jin, W., Liang, Z., Abomohra, A. E. F., Zhou, X., Tu, R., & Han, S. (2017). Abundance and diversity of ammonia-oxidizing archaea in a biological aerated filter process. *Annals of Microbiology, 67*(6), 405–416.

Das Gupta, A., Sarkar, S., Singh, J., Saha, T., & Sil, K. A. (2016). Nitrogen dynamics of the aquatic system is an important driving force for efficient sewage purification in single pond natural treatment wetlands at East Kolkata Wetland. *Chemosphere 164*, 576–584.

DeBusk, W. F. (2003). *Nitrogen cycling in wetlands* (pp. 1–3). University of Florida IFAS Ext (SL171).

Denancé, N., Szurek, B., & Noël, L. D. (2014). Emerging functions of nodulin-like proteins in non-modulating plant species. *Plant and Cell Physics, 3*, 469–474.

Fink, P., Pflitsch, C., & Marin, K. (2011). Dietary essential amino acids affect the reproduction of the keystone Herbivore Daphnia pulex. *PLoS ONE, 6*(12), 1–28498.

Galloway, N. J., Aber, D. J., Erisman, J. W., Seitzinger, P. S., Howarth, W. R., Cowling, E. B., & Cosby, J. B. (2003). The Nit Cscd. *BioScience, 53*, 341–356.

Gangagni Rao, A., SasiKanth Reddy, T., Surya Prakash, S., Vanajakshi, J., Joseph, J., Sarma, P. N. (2007). pH regulation of alkaline wastewater with carbon dioxide: A case study of treatment of brewery wastewater in UASB reactor coupled with absorber. *Bioresource Technology* (98), 2131–2136.

Gao, J., Luo, X., Wu, G., Li, T., & Peng, Y. (2014). Abundance and diversity based on amoA genes of ammonia-oxidizing archaea and bacteria in ten wastewater treatment systems. *Applied Microbiology and Biotechnology, 98*(7), 3339–3354.

Gwak, J. H., Jung, M. Y., Hong, H., Kim, J. G., Quan, Z. X., Reinfelder, J. R., Spasov, E., Neufeld, J. D., Wagner, M., & Rhee, S. K. (2020). Archaeal nitrification is constrained by copper complexation with organic matter in municipal wastewater treatment plants. *The ISME Journal, 14*(2), 335–346.

Hoffman, B. M., Lukoyanov, D., Yang, Z. Y., Dean, D. R., & Seefeldt, L. C. (2014). Mechanism of nitrogen fixation by nitrogenase: The next stage. *Chemical Reviews, 8*, 4041–4062.

Hoffman, J. R., & Falvo, M. J. (2004). Protein–which is best? *Journal of Science and Medicine in Sport*, (3), 118.

Hoghooghi, N., Pippin, J. S., Meyer, B. K., Hodges, J. B., & Bledsoe, B. P. (2021). Frontiers in assessing septic systems vulnerability in coastal Georgia, USA: Mod app and mngmt impl. *PloS one*, (8), e0256606.

Kampman, C., Temmink, H., Hendrickx, T. L. G., Zeeman, G., & Buisman, C. J. N. (2014). Enrichment of denitrifying methanotrophic bacteria from municipal wastewater sludge in a membrane bioreactor at 20 °C. *Journal of Hazardous Materials, 274*, 428–435.

Kraft, B., Tegetmeyer, H. E., Sharma, R., Klotz, M. G., Ferdelman, T. G., Hettich, R. L., Geelhoed, J. S., & Strous, M. (2014). The environmental controls that govern the end product of bacterial nitrate respiration. *Science, 345*(6197), 676–679.

Landner, L. (2013). Sources of nitrogen as a water pollutant: industrial wastewater. Proc of the Conf on Nit As a Wat Pt: 55–65

Lee, C.-G., Fletcher, T., & Sun, G. (2009). Nitrogen removal in constructed wetland systems. *Engineering in Life Sciences* (9), 11–22.

Lei, M., Kai, W., Shuying, W., Rulong, Z., Baikun, L., Yongzhen, P., & Dongchen, W. (2014). Advanced nitrogen removal from landfill leachate using real-time controlled three-stage sequence batch reactor (SBR) system. *Bioresource Technology, 159*, 258–326.

Lei, M., Gangqing, Y., Tao, T., & Yongzhen, P. (2019). Recent advances in nitrogen removal from landfill leachate using biological treatments–A review. *Journal of Environmental Management, 235*, 178–185.

Lie, Z., Huang, W., Liu, X., Zhou, G., Yan, J., Li, Y., Huang, C., Wu, T., Fang, X., Zhao, M., Liu, S., Chu, G., Kadowaki, K., Pan, X., & Liu, J. (2021). Warming leads to more closed nitrogen cycling in nitrogen-rich tropical forests. *Global Change Biology, 27*(3), 664–674.

Liu, X. Y., Koba, K., Makabe, A., & Liu, C. Q. (2014). Nitrate dynamics in natural plants: Insights based on the concentration and natural isotope abundances of tissue nitrate. *Front in Plant Science, 5*, 355.

Mariotti, F., Tomé, D., & Mirand, P. P. (2008). Converting nitrogen into protein—beyond 6.25 and Jones' Factors. *Critical Reviews in Food Science and Nutrition*, (48), 2, 177–184.

Poh, P. E., Gouwanda, D., Mohan, Y., Gopalai, A. A., & Tan, H. M. (2016). Optimization of wastewater anaerobic digestion using mechanistic and meta-heuristic methods: current limitations and future opportunities. *Water Conservation Science and Engineering, 1*, 1–20.

Rajta, A., Bhatia, R., Setia, H., & Pathania, P. (2020). Role of heterotrophic aerobic denitrifying bacteria in nitrate removal from wastewater. *Journal of Applied Microbiology, 128*, 1261–1278.

Roy, A. H. (2007). Fertilizers and food production. In J. A. Kent (Ed.), *Kent and Riegel's handbook of industrial chemistry and biotechnology* (pp. 1111–1156). Springer.

Roy, A., Aftabuddin M. K. M. S. (2018). Conflict around wetland use in West Bengal. Ind wat port: 1–1

Shapley, P. (2009). *Wetlands and chemistry, US EPA* (p. 1)

Sliekers, A. O., Derwort, N., Campos Gomez, J. L., Strous, M., Kuenen, J. G., & Jetten, M. S. M. (2002). Completely autotrophic nitrogen removal over nitrite in one single reactor. *Water Research, 36*, 2475–2482.

Sponseller, R. A., Gundale, M. J., Futter, M., et al. (2016). Nitrogen dynamics in managed boreal forests: Recent advances and future research directions. *Ambio, 45*, 175–187.

Tang, C.-J., Zheng, P., Chen, T.-T., Zhang, J.-Q., Mahmood, Q., Ding, S., Chen, X.-G., Chen, J.-W., & Wu, D.-T. (2011). Enhanced nitrogen removal from pharmaceutical wastewater using SBA-ANAMMOX process. *Water Research* (45), 201–210.

United States Environmental Protection (2021), The Sources and Solutions: Wastewater, Nutrient Pollution, US Env ProtAgcy: 1–1

van Puijenbroek, P. J. T. M., Beusen, A. H. W., & Bouwman, A. F. (2019). Global nitrogen and phosphorus in urban waste water based on the Shared Socio-economic pathways. *Journal of Environmental Management, 231*, 446–456.

Ward, B. B. (2003). Significance of anaerobic ammonium oxidation in the ocean. *Trends in Microbiology, 11*, 408–410.

Washington State Department of Health. (2014). How Nitrogen from Septic Systems Can Harm Water Quality, WS DOH: 1–2

Watford, M., & Wu, G. (2018). Protein. *Advances in Nutrition*, (5), 651–653.

Wiesmann, U. (1994). Biological nitrogen removal from wastewater. In U. Wiesmann (Ed.), *Biotechnics/Wastewater* (pp. 113–154). Springer.

Wolf, K. L., Ahn, C., & Noe, G. B. (2011). Development of soil properties and nitrogen cyclingin created wetlands. *Wetlands* (31), 699–712.

Wong, C., Barton, G., & Barfor, J. (2003). The nitrogen cycle and its application in wastewater treatment. *Hndbk Wat and Wstwatr Microbiol* 427–439.

Yin, Z., Xie, L., & Zhou, Q. (2015). Effects of sulfide on the integration of denitrification with anaerobic digestion. *Journal of Bioscience and Bioengineering, 120*(4), 426–431.

Zhou, N. Q., Yang, W. J., & Wang, Y. (2010a). Nitrogen cycle characteristics in wetlands and its influence to natural environment. *International Conference on Automobile and Mechanical Engineering*, 2105–2108.

Zhou, N. Q., Yang, W. J., & Wang, Y. (2010b). Nitrogen cycle characteristics in wetlands and its influence to natural environment. *International Conference on Automobile and Mechanical Engineering*, 2105–2108.

Zhou, N., Yang, W., & Wang, Y. (2010c). Nitrogen cycle characteristics in wetlands and its influence to natural environment. MACE 2010c

Zhou, N. Q., Yang, W. J., & Wang, Y. (2012). Nitrogen cycle characteristics in wetland and its influence to natural environment. In *IEEE 2010 International Conference on Mechanic Automation and Control Engineering (MACE)*. Wuhan

Recent Advances in the Biochemistry and Molecular Biology of Ammonia-Oxidizing Bacteria

Ayush Jha, Vishal Dutta, Saipriya Dhawan, Karambir Singh, Pitambri Thakur, and Gunjan Mukherjee

Abstract Ammonia deposition in aquatic habitats has a severe influence on water quality and human health, primarily to the excessive proliferation of toxic algae. Ammonia prevents oxygen from reaching the fish's gills, resulting in death. Furthermore, humans can also become ammonia victims indirectly via eating the polluted seafood and water. As strict regulatory guidelines demand all these industries should treat wastewater prior being disposing it to the surrounding environment. As a result, the elimination of ammonia from wastewater is necessarily important. Environmental regulations and strict discharge limitations are prompting the industries to explore out for more sustainable and cost-effective approaches for elimination of NH_3 from wastewater. Ammonia may be eliminated through wastewater via a variety of biological, chemical, and physical techniques. The microbiological approach is the most conventional and cost-effective method of ammonia removal. Technologies associated with the elimination of ammonia such as anaerobic ammonia oxidation and full ammonia oxidation are relatively recent. These technologies aid in the development of technologies for the rapid and cost-effective removal of organic pollutants like ammonia from wastewater. This chapter comprehensively represents various sources resulting in addition of organic pollutants in wastewater, and recent advances in the biochemistry and molecular biology of ammonia-oxidizing bacteria, and their influence on wastewater ammonia oxidation.

Keywords Ammonia oxidizers · Nitrite oxidizing bacteria · Anaerobic ammonia oxidation

---

A. Jha · P. Thakur · G. Mukherjee (✉)
University Institute of Biotechnology, Chandigarh University, Mohali, Punjab, India
e-mail: gunjanmukherjee@gmail.com

V. Dutta
School of Advanced Chemical Sciences, Shoolini University, Solan, HP, India

S. Dhawan
Ms Swaminathan School of Agriculture, Shoolini University, Solan, HP, India

K. Singh
Special Center for Nanoscience, Jawaharlal Nehru University, New Delhi 110067, India

M. P. Shah (ed.), *Anammox Technology in Industrial Wastewater Treatment*,
https://doi.org/10.1007/978-981-99-3459-1_9

1 Introduction

Wastewater reduction and contaminants elimination specifically nitrogen containing wastes are prime concern globally. The most prevalent nitrogen-based contaminant introduced into aquatic habitats is ammonia. In natural conditions, the quantity of ammonia in groundwater and surface water is less than 0.2 mg/L. Discharge of urban and industrial wastewater, as well as agricultural activities, has dramatically increased ammonia concentrations in groundwater and surface water (Bartlett et al., 2008). Ammonia deposition in aquatic habitats has a severe influence on water quality and human health, primarily to the excessive proliferation of toxic algae. Excessive cyanobacteria and green algae growth and aging deplete dissolved oxygen and trigger the production of cyanotoxins, resulting in the expiry of aquatic life (Morris et al., 2019). Ammonia prevents oxygen from reaching the fish's gills, resulting in death. Furthermore, humans can also become ammonia victims indirectly via eating the polluted seafood and water. Ammonia concentrations in municipal wastewater are minimal (around 35–45 mg/L), while anaerobic digester effluent includes large levels of ammonia (EPA et al., 2013). In addition to urban wastewater, NH_3-containing nitrogen compounds are released into natural streams by industrial and agricultural operations. Water containing ammonia is an equilibrium formed between ionized (NH^+_4) and un-ionized (NH_3) types of ammonia. The temperature and pH of the water primarily affect the concentration of both these ammonia forms (Qing et al., 2012). Un-ionized ammonia is a predominantly toxic form of ammonia because it can pass through the epithelial tissues of aquatic species more readily than ionized (NH^+_4). Under exceptional conditions, ammonium ions have also been linked to substantial toxicity. Un-ionized ammonia is more abundant at higher pH than at lower pH (Atkins et al., 1973). Nitrite ions can be converted to free nitrous acid when the pH is reduced (Liu et al., 2019a). Nitrate at high concentrations has a considerable harmful effect on harm human health. The concentration of ammonia in municipal solid waste streams varies significantly. Industrial waste streams, fertilizer plants, and landfill leachate typically have very high ammonia content (Cardemil et al., 2010). As strict regulatory guidelines demand all these industries should treat wastewater prior to being disposing it to the surrounding environment. As a result, the elimination of ammonia from wastewater is necessarily important (Park et al., 2018). Environmental regulations and strict discharge limitations are prompting the industries to explore more sustainable and cost-effective approaches for the elimination of NH_3 from wastewater. By utilizing various techniques such as physical, biological, and chemical techniques ammonia can be eliminated from wastewater. Ultrafiltration, nanofiltration, reverse osmosis and ion exchange with natural minerals, ozone oxidation, chemical precipitation, and iron stripping are all physiochemical techniques used for the elimination of ammonia from wastewater. These systems each have their own set of advantages and limitations, with the most significant negative being the high operational costs and specialized personnel requirements (Lei et al., 2012). The high operational cost emphasizes the need of adopting biological methods to eliminate ammonia from wastewater since they are both cost-effective and environmentally benign. The microbiological approach is the most conventional and easy ammonia removal method. Microbes are used in the traditional microbiological approach for nitrification and de-nitrification (Yao et al., 2019). Nitrosomonas and Nitrobacter are examples of nitrifying bacteria helping in nitrous and nitric oxide synthesis from ammonia (Daims et al., 2015). Technologies associated with elimination of ammonia such as anaerobic ammonia oxidation and full ammonia oxidation are relatively recent. Conversion of ammonia to nitrogen with the help of microorganisms via anammox process (anaerobic ammonium oxidation) is low-cost and ecofriendly technique. Comammox comprises ammonia nitrification to nitrate and nitrite and it is a two-step chemolithoautotrophic bacterium process. The most frequent bacteria explored in the comammox process are *Nitrospira* species (Zhang et al., 2008).

The present chapter explores the ANAMMOX and COMAMMOX processes, as well as the microorganisms involved and their biochemistry and molecular biological advancements. These technologies may aid in the development of technologies for the rapid and cost-effective removal of organic pollutants like ammonia from wastewater (Ali & Satoshi, 2015). This chapter comprehensively represents various sources resulting, in addition, organic pollutants in wastewater, the structure of anammox and comammox bacteria, the physiological process associated with anammox and comammox bacteria, the similarity of anammox process with the nitrogen cycle, the interaction of anammox and comammox and its influence on wastewater ammonia oxidation. The prime objective of this chapter is to offer a detailed overview of the evolution of the anammox and comammox ammonia elimination methods, as well as efforts for these two approaches for water mitigation (Ali et al., 2015).

2 Nitrogen Cycle

Nitrogen (N) belongs to group 5B and has oxidation states ranging from −3 to + 5. The nitrogen atom unites with hydrogen, oxygen, or other nitrogen atoms in each oxidation state. In this manner, each oxidation state has at least one distinct inorganic molecule. Moreover, some nitrogen-based compounds are even more stable than others because the oxidation state of N in a particular environment is governed by high activation energy of the N-compounds, rather than thermodynamic equilibrium, all oxidation states are conceivable in aqueous systems. The majority of nitrogen on the planet is in solid form (rocks). However, the most essential nitrogen source available to organisms is the dinitrogen gas in the atmosphere (78.5%). Nitrogen is essential for life. A live organism's general chemical formula is $CH_2O_{0.5}N_{0.15}$. The nitrogen cycle (the turnover of nitrogen compounds in the biosphere) is comprised of five catabolic mechanisms (nitrification, nitratification, denitrification, dissimilatory nitrate reduction, and anammox), three anabolic processes (ammonium uptake, assimilatory nitrate reduction, and nitrogen fixation), and ammonification (ammonium uptake, assimilatory nitrate reduction, and nitrogen fixation) (an essential product of the biological food chain) (Li & Ji-Dong, 2011). Table 1 summarizes the most essential nitrogen cycle enzymes and the processes that they catalyze. The main objective of nitrogen cycle microbiology in the earlier years was to better understand and enhance fertilizer performance in agriculture. The potential of nitrifiers and denitrifiers for nutrient removal from wastewater was not widely recognized until the 1960s when studies were focused on improving denitrification from wastewater (Hong et al., 2011). The significance of nitrification and denitrification in the formation of these compounds was, however, the focus of "environment-driven" N-cycle research in the 1980s when the contribution of nitrogen oxides in the atmosphere to ozone depletion and global warming was reviewed (Baolan et al., 2012). However, the last decade has demonstrated that our understanding of the microbial nitrogen cycle and its key participants are far from comprehensive. Current researches demonstrate that the microbial world contains a huge variety and metabolic potential for nitrogen conversions, which is currently a hot topic. Remarkable findings such as reduction of nitrate to dinitrogen gas via foraminifera, oxidation of nitrite phototrophs, anaerobic ammonium oxidation (anammox), hyperthermophilic, nitrogen fixation via methane-producing bacteria, genome sequencing of several N-cycle organisms, nitrite-dependent and anaerobic methane oxidation (N-DAMO) (Li et al., 2020). In addition, new sequencing technologies and the modification of molecular methodologies have shown insights the great majority of functional microbial diversity in the environment still holds. Increased fossil fuel combustion and high nitrogen demand in agriculture and industry, on the other hand, imply that humankind is driving to modify the global N-cycle at a breakneck pace. Vast quantities of anthropogenic nitrogen are discharged into the environment, causing a cascade of issues such as increasing freshwater nitrate levels and nitrous oxide generation, which may contribute to the greenhouse effect (Dang et al., 2010). To comprehend and eventually prevent the deleterious effect of nitrogen pollution, a better understanding of the bacteria responsible for nitrogen transformations is critically required (Wang et al., 2012). Various reviews on microbial nitrogen cycle have been published over the last 10 years, interestingly the majority of them focused on microbial nitrogen cycle for organic pollutant removal. From 2011 to 2021, a search of the "Scopus database" using the keywords "microbial nitrogen cycle for environmental remediation" found approximately 1,16,000 publications (Fig. 1a). The anammox process has advanced from a relatively unexplored component of the biological nitrogen cycle to public practice since its discovery in 1995. Anammox and comammox bacteria have emerged as prominent participants in the global nitrogen cycle over the last several years. A survey based on the anammox process illustrates that from 2011 to 2021 using the keywords "anammox and comammox process for organic pollutant elimination" displays near about 7,320 publications out of which 5,210 are concerning the anammox process (Fig. 1b) and 2,110 are reported for the comammox process (Fig. 1c). This portrays the keen concern over the environmental remediation using these microbial processes.

Table 1 Essential nitrogen cycle enzymes and the processes that they catalyze (Wang et al., 2012)

Steps involved	Processes/Reaction	Cell potential	Site
Ammonia monooxygenase	$NH_2OH + H_2O$	0.73 V	Transmembrane
Hydroxylamine oxidoreductase	$NO_2^- + 5H^+ + 4e^-$	−0.06 V	Periplasm
Nitrite oxidoreductase	$NO_3^- + 2H^+ + 2e^-$	−0.43 V	Membrane
Hydrazine hydrolase	$N_2H_4 + H_2O$	0.34 V	Anammoxosome membrane
Hydrazine oxidoreductase	$N_2 + 4H^+ + 4e^-$	−0.75 V	Annammoxosome
Nitrate reductase	$NO_2^- + H_2O$	0.43 V	Cytoplasmic or periplasmic membrane
Nitrite reductase	$NO + H_2O$	0.34 V	Periplasm
Nitric oxide reductase	$N_2O + H_2O$	1.17 V	Transmembrane
Nitrous oxide reductase	$N_2 + H_2O$	1.36 V	Periplasm
Dissimilatory nitrite reductase	$NH_4^+ + 2H_2O$	0.75 V	Cytoplasm

3 Growth and Structure of Bacteria

3.1 Growth of Bacteria

Bacteria engaged in anaerobic ammonia oxidation may be found almost everywhere. Planctomycetes are the phylum in which these bacteria reside. Only 10 of the 19 anammox species discovered thus far, belong to five taxa, could be enriched and are grown at a pH range of 6.8–8.5, which is neither acidic nor basic (Seung et al., 2005). Anammox bacteria could well be found in both freshwater and marine habitats. These can also be found in treating wastewater reactors and agricultural land (Hsia et al., 2008). Recent research has found that these bacteria may be enhanced more quickly in groundwater than in deionized water. Anammox bacteria are chemolithoautotrophic, meaning that they may use metals and trace elements found in groundwater (Chen et al., 2016). They are autotrophic bacteria that grow slowly, with replication times ranging from 10 to 25 days. Anammox bacteria develop at different temperatures in both marine and freshwater (Thu et al., 1996). These bacteria grow best at a temperature range of 15–30 °C in marine waters, while in freshwater, they exhibit maximal activity at 38 °C, generally ranging from 28 to 38 °C. As anammox bacteria are not culturable, procedures like qPCR and FISH are employed to detect and characterize their quantity. According to studies, the population of anammox fluctuates in sediment samples from different locations and is influenced by pollution. The abundance

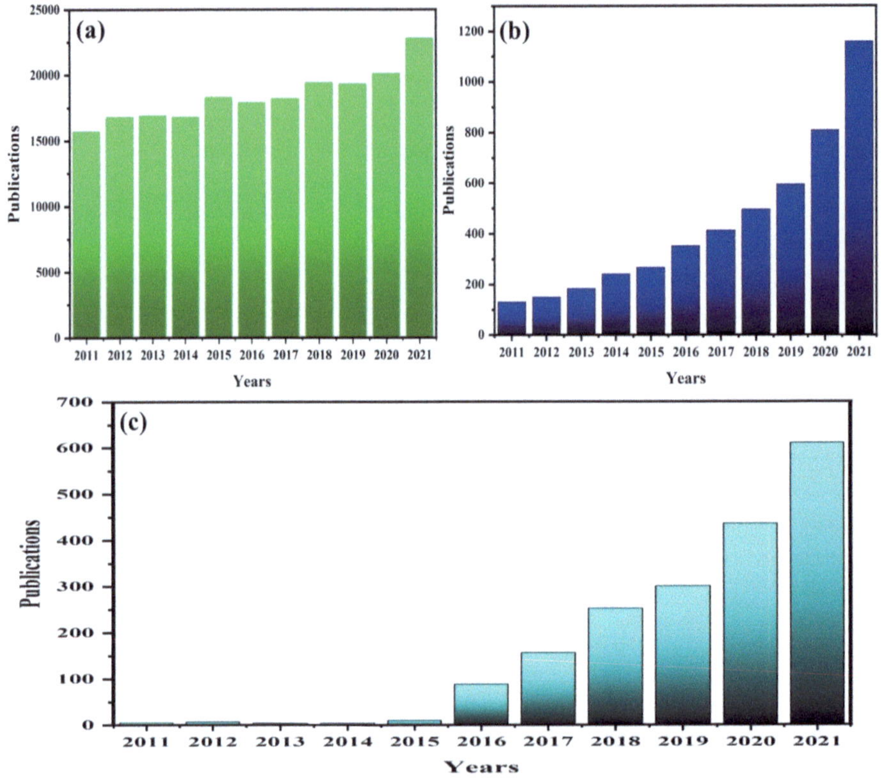

Fig. 1 **a** From 2011 to 2021, a search of the "Scopus database" using the keywords "microbial nitrogen cycle for environmental remediation" found approximately 1,16,000 publications **b,** Anammox process illustrates that from 2011 to 2021 using the keywords "anammox and comammox process for organic pollutant elimination" displays near about 7,320 publications out of which 5,210 are concerning anammox process **c** 2,110 are reported for comammox process (Leenen et al., 1996; Takeshima et al., 1993)

values for the anammox-specific gene *hzo* and 16 s rDNA obtained by qPCR vary from 1.20×10^4 to 6.0×10^6 copies/g of sediment (Leenen et al., 1996). As there may be many copies of the hzo gene in each bacterium cell, the abundance determined using hzogene may not be precise, but it does cover all anammox species.

Anammox co-exists with methane-oxidizing bacteria in paddy fields, according to reports (Takeshima et al., 1993). Anoxic conditions emerge in paddy fields as a result of waterlogging, which is ideal for anammox bacteria. Anammox has been recorded from both mesophilic and thermophilic environments, such as oil fields in China (Sumino et al., 1992).

Comammox is a process in which a single organism completes the oxidation of ammonia to nitrate. When compared to independent nitrification stages, it produces better growth yields (Sumino et al., 1992). Comammox are found in abundance in the environment. These can be found in both terrestrial and aquatic habitats.

These can be found in agricultural soils, such as rice paddy fields, as well as forest soils. Freshwater habitats, brackish lake sediments, and artificial systems like wastewater treatment facilities and drinking water treatment systems are all used to cultivate them. Comammox are oligotrophic organisms that convert ammonia to nitrite and nitrate in low-nutrient environments (Hu et al., 2014). Comammox dominated anaerobic ammonia oxidation (AOB) in drinking water treatment facilities and aided ammonia oxidation. In comparison to anaerobic ammonia oxidation (AOB), comammox bacteria have a more effective carbon fixation mechanism and so require a low amount of organic carbon (Østgaard et al., 1993). In the mainline nitrification reactor with little dissolved oxygen, they are the primary ammonia oxidizers. They thrive in low-dissolved oxygen or partly toxic environments (Magrí et al., 2012). Phototrophic bacteria are cultured in dark circumstances to restrict their proliferation. These bacteria can thrive in a wide range of temperatures, from 20 °C to 55 °C, and at pH levels ranging from 6 to 9. Many designed systems have been found to contain *Comammox Nitrospira* species (Isaka et al., 2017). It was also discovered to be the dominant ammonia-oxidizing community in a municipal wastewater treatment plant's nitrification reactor. Under severely oligotrophic conditions, physiological data like ammonia oxidation kinetics, metabolic flexibility, and comparative genomic investigation demonstrated that comammox organisms functionally outcompete other classical nitrifiers (Li et al., 2010). The anammox and comammox processes are distinct, and their differentiation based on their performance is listed in Table 2.

Table 2 The anammox and comammox processes are distinct and their differentiation is based on their performance (Li et al., 2010)

	Anammox (Anaerobic-Ammonia Oxidation)	Comammox (Complete-Oxidation Ammonia)	Reference
Structure	Anammox are gram-negative (Coccid bacteria)	Comammox are Spiral-shaped, aerobic with a flagellum	Li et al. (2010)
Growth period	11–22 days	Up to 2 weeks	Li et al. (2010)
Species found	*Brocadia, Kuenenia, Anammoglobus, Jettenia, Scalindua*	*Ca. Nitrospira inopinata, Ca. Nitrospira japonica, Ca. Nitrospira nitrosoa, Ca. Nitrospira nitrificans*	Ali et al. (2013)
Process	Partial nitrification $NH_4^+ + NO_2^- \rightarrow N_2 + 2H_2O$	Complete Nitrification $NH_4^+ + NO_2^- \rightarrow NO_3^-$	van Teeseling et al. (2014)
Advantages	CO_2 emission and low sludge production	Higher yield growth, Complete nitrification	Viancelli et al. (2011)
Application	Wastewater treatment and industrial usage	Drinking water treatment	Kartal et al. (2007)

3.2 Structure of Bacteria

Anammox bacteria have a distinctive shape and biology. In anammox bacteria, the cytoplasm is divided into three compartments by a single bilayer membrane (Li et al., 2018). Anammoxosome, riboplasm, and paryphoplasm are the three bound compartmental cells from the inside out. Anammoxosomes are specialized organelles in anammox bacteria that include a single bilayer membrane and provide a site for metabolism. ATP synthesis in the riboplasm is aided by a proton motive force generated across the anammoxosome membrane during an anammox reaction (Ali et al., 2020). Proton diffusion and transitional toxicity are prevented by the anammoxosome membrane (Mardanov et al., 2019). Ladderane lipids, which are organic molecules having two or more joined cyclobutane rings, are found naturally in bacterial membranes. Anammox could not be cultivated as a pure isolate; therefore, genomic information was collected in parts and pieces. *Candidatus Jettina* genome was recently determined from metagenomic analysis of anammox granules (Swathi Desireddy et al., 2018). Anammox bacteria are supplemented in bioreactors such as membrane reactor, sequencing batch reactor, moving bed biofilm reactor, fixed-film reactors, up-flow anaerobic sludge bed, fluidized-bed, fixed-bed, rotating biological contactor, and granular sludge-based reactor (Swathi Desireddy et al., 2017). The combining of the content of the reactor was simple in a granular sludge-based reactor, and the volumetric loading rate was greater. Biomass retention is effective due to the anammox bacteria's slower growth, resulting in the natural formation of biofilm (Arrojo et al., 2006). Anammox bacteria grow in granular form as consortia with other microorganisms. The anammox bacterial growth can be visible in the center of the granules, while the lesser percentage of ammonia-oxidizing bacteria (AOB) growth can be seen on the granules' outer shell. These granules range in color from brilliant red, carmine, calamine, to pale red or black. Anammox bacteria grow in reddish-brown due to the high quantity of cytochrome, and the color fades to black as the amount of Heme C is reduced. Anammox bacteria take a long time to grow. The cells multiply every 12–20 days. In situ circumstances minimize the doubling period, allowing these microorganisms to flourish in low-nutrient environments. Anammox are oxygen-sensitive obligate anaerobes (Mulder et al., 1995). If the oxygen level is higher than 2 μM, their metabolism is extracellular suppressed. Most of these are chemolithoautotrophic, implying they can use nitrate as an electron acceptor to oxidize ammonia to nitrogen gas.

Scanning electron microscope (SEM) image of pure cultures of comammox bacteria such as *Nitrospira inopinata*, ca. *N. uzonensis*, was revealed to be somewhat curved or spiral-shaped, with a few being round-shaped in size with 0.2–0.7 μm in width and 0.7–2 μm in length. A multi-layered envelope, consisting of a cytoplasmic membrane and an outer cellular membrane, surrounds these cells. As deposits, the membrane is made up of storage molecules such as glycogen and polyphosphate. Rotating biological contactors, sequencing batch reactor, and moving bed biofilm reactor all include a high concentration of Comammox bacteria. *Nitrospira inopinata* has an abundance of autotrophic nutritional medium that is enriched in ammonia.

They are chemolithoautotrophic in most occurrences. These microorganisms grow in biofilms or a floating growth environment. These biofilms form an extracellular polymeric surface network that comprises 60–90% of the complete organic matter (Development of a Fixed-Bed Anammox Reactor with High Treatment Potential, 2013).

4 Reaction Mechanism and Enzymes Involved

4.1 Anammox Reaction Mechanism and Enzyme Involved

The potential to anaerobically oxidize ammonia is a rare metabolic trait of anammox bacteria. As for the last electron, they use the nitrite electron as an acceptor in ammonium oxidation, resulting in the production of dinitrogen gas as a byproduct product. The essential agents in achieving this requirement are bicarbonate or carbon dioxide. Anammox bacteria have carbon needs and so they are classified as autotrophic microorganisms. Anammox bacteria are a type of bacterium that falls under the chemolithoautotrophic bacteria that are obligately anaerobic. In general, microbes use four different methods to help fix Carbon dioxide from the air. The Calvin cycle, the 3-hydroxypropionate pathway, the reverse citric acid cycle, and the acetyl-coenzyme A (acetyl-CoA) pathway are among them. In order to fix ambient CO_2, Anammox bacteria may use the Calvin cycle or the Acetyl-CoA pathway. Anammox bacteria have a membrane-bound organelle called an anammoxosome, which is where the anaerobic process reactions occur. Antibodies were used to confirm that this adds to the evidence that the anammoxosome is only found in anammox bacteria (Fig. 2a). In this case, the proton motive force required for ATP production is generated and maintained by the anammoxosome membrane. Anammox metabolism consists of a sequence of enzyme-catalyzed processes. Key enzymes involved in the anammox process are found inside the anammoxosome. Anammox metabolism consists of a sequence of enzyme-catalyzed processes. The equation depicts the total anammox reaction or energy-generating process in an anammox cell (Okamoto et al., 2013; Wang et al., 2016).

$$\begin{aligned} &NH_4^+ + 1.32NO_2^- + 0.066HCO_3^- + 0.13H^+ \rightarrow \\ &1.02N_2 + 0.26NO_3^- + 0.066CH_2O_{0.5}N_{0.15} + 2 \cdot 03H_2O \end{aligned} \tag{1}$$

The following equation specifies that the overall anammox reaction is a mixture of two distinct reactions (Eqs. 1 and 2). As shown in Eq. 1, ammonium and nitrite are transformed into nitrogen gas, and the free energy associated with this reaction is significant (G = 375 kJ/mol), which could be used in the development of anammox bacteria that rely on inorganic substances for energy. Anammox bacteria use hydrocarbons as their only carbon source for biomass formation, making them autotrophic. The oxidation of nitrite to nitrate produces the reducing equivalents need for biomass

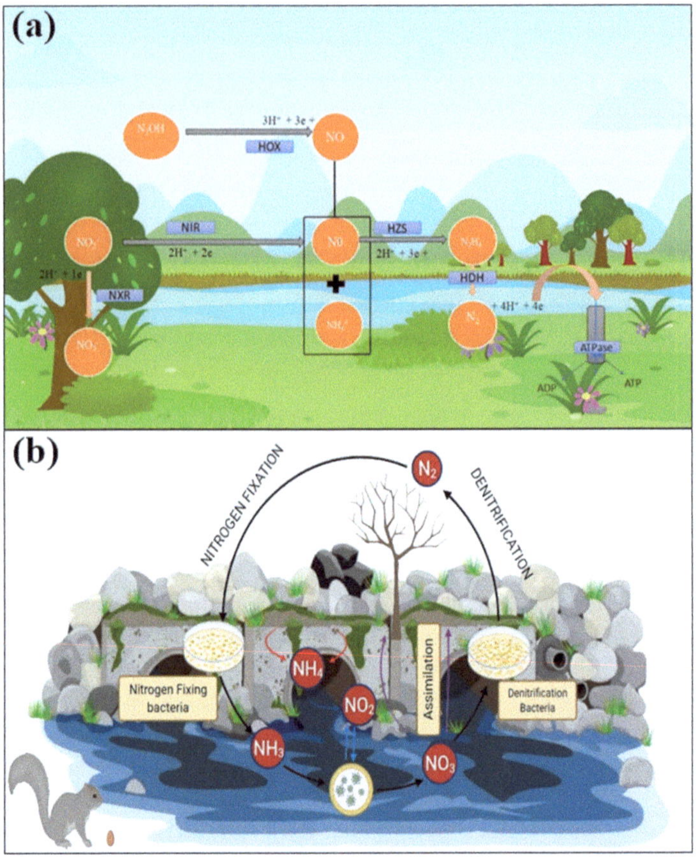

Fig. 2 **a** Schematic representation of anammoxosome catabolism. **b** Nitrogen cycle in respect to wastewater mitigation (Shah, 2021)

synthesis (Eq. 2) (Li & Sung, 2015).

$$NH_4^+ + NO_2^- \rightarrow N_2 + 2H_2O + 0.26NO_2^- + 0.066HCO_3^- \rightarrow 0.26NO_3^- + 0.066CH_2O_{O.5}N_{0.15} \quad (2)$$

4.2 *Comammox Reaction Mechanism and Enzyme Involved*

Winogradsky discovered in 1890 that nitrogen fixation is a dual-step process involving ammonia-oxidizing and nitrite-oxidizing microorganisms. The nitrite-oxidizing bacteria (NOB) and the ammonia-oxidizing bacteria (AOB) are not closely linked phylogenetically. Microbial division of labor resulted in the highest generation of ATP, according to the best metabolic pathway design kinetic theory. The growth yield is increased by shortening the pathway. The bacteria's increased growth yield would benefit from growing them in clonal clusters in the form of biofilms. The newly found method of complete ammonia oxidation demonstrates that a single organism may convert ammonia to nitrite and nitrate oxidation. These bacteria are employed in biological technology design and process because of their distinct metabolic functionality and diversified existence. Complete nitrification produces the same amount of energy as the two-step processes of ammonia to nitrite and nitrite to nitrate oxidation (275 kJ/mole and 74 kJ/mole, respectively). The organisms involve in complete nitrification had higher growth and energy yield than incomplete ammonia-oxidizing bacteria and nitrite-oxidizing bacteria, which could be related to complete nitrifiers' competition. The comammox species may thrive in extremely lithotrophic environments with limited nutrition and DO availability. It was discovered that comammox outweighed AOB in drinking water systems. The traditional nitrification process via AOB and NOB is aerobic oxidation, with NOB enzymes inactive in anoxic conditions. Comammox thrives in low anoxic environments, making nitrification more efficient. Because the Comammox enzyme has a larger affinity for ammonia than the AOB enzymes, it overtakes nitrification and so takes control of the system (Egli et al., 2001; Swathi Desireddy et al., 2020).

5 Anammox Bacteria Process in Wastewater Mitigation, Resemblance with the Nitrogen Cycle

Anthropogenic sources lead to nitrogen emissions in a variety of ways as the world's population grows. Excess nitrogen released into natural streams causes concerns such as soil acidification and eutrophication, among other things. The legislation and the communities have imposed higher effluent standards for nutrient wastewater in the previous decade because ammonia is a dangerous nitrogen-containing compound with several environmental and health consequences, various biological techniques and treatment methods are used to remove ammonia from wastewater (Rossi et al., 2020). Anammox bacteria are stringent chemolithoautotrophs, meaning they make energy by using inorganic chemicals like ammonia and nitrite as electron donors and acceptors (Liu et al., 2019b). Other electron donors, including organic compounds and amines, can be used by some anammox bacteria. Anammox genera like *Candidatus Anammoxoglobus propionicus* and *Candidatus Brocadia fulgida* exhibit a stronger preference for organic compounds like acetate and propionate, demonstrating the metabolic plasticity of anammox bacteria. Anammox bacteria can also use inorganic metal oxides like Mn (IV) and Fe (III) as electrophiles instead of nitrite (Jiang et al., 2018). In a dynamic mixed microbial environment, these organisms' species-specific metabolic differences give them an adaptive edge. Anammox bacteria relate to the Planctomycetes phylum and have a sluggish metabolism.

Anammox was previously thought to oxidize ammonia using nitrite as an electrophile (Shah, 2020). Later, it was shown that anammox could utilize other electron acceptors besides nitrite, confirming their metabolic plasticity. It was also discovered that organic acids (propionate) compete with ammonium for the electrophile nitrite. Anammox sludge was unable to oxidize propionate in the absence of nitrite or nitrate, indicating that propionate and ammonium are unrelated. The name *Anammoxoglobus propionicus* was proposed in a later investigation (Shah, 2021). The anammox technique has a wide range of applications for the removal of nitrogen as well as organic compounds from wastewater, according to the newly identified anammox strain. Iron might potentially be oxidized using nitrite as an electron receiver, according to the theory.

Anthropogenic activities impact nitrogen emissions in a variety of ways as the world's population grows. Excess nitrogen releases into natural streams cause concerns such as soil acidification and eutrophication, among other things (Shah Maulin, 2021; Costa et al., 2006) . The law and the community have imposed higher effluent requirements for nutrient wastewater in the previous decade. As ammonia is a toxic nitrogen-containing compound with several environmental and health consequences, a variety of biological techniques and treatment systems are used to remove ammonia from wastewater. Microalgae-based assimilation, non-thermal plasma combined with natural minerals such as zeolites, membrane technology employing microbial cells, ammonia stripping, and heterotrophic ammonia oxidation are only a few examples. The relevance of using anammox bacteria in wastewater treatment over these methods is demonstrated by a comparison of these approaches with the anammox process. Reported by Rossi et al. (2020) certain microalgae genera used in wastewater ammonia treatment are extremely sensitive to variations in free ammonia concentrations in the persistence medium, resulting in irreversible inhibition of the photosynthetic manner that meets the microalgae's oxygen requirements. Anammox bacteria, on the other hand, have a higher tolerance for free ammonia than microalgae used in ammonia treatment. In addition, unlike microalgae, the inhibition induced on anammox bacteria by free ammonia is permanent. However, as per previous findings, the ammonia removal effectiveness of non-thermal plasma linked with zeolites decreases as the ammonia content of effluent wastewater rises, limiting the use of this technique in ammonia-rich wastewater. To meet the extra energy requirements, the energy generated by microbial cells employing membrane technology is used in wastewater treatment systems. However, this comes at a high expense of operating. The first and most common biological technique for removing nitrogen species from wastewater was nitrification/denitrification, which involves converting ammonia to nitrate and then reducing the nitrate to diatomic nitrogen. However, when wastewater includes just a tiny amount of compounds of biodegradable carbon as well as aeration systems, this can become a significant cost factor. Meanwhile, the anammox process has achieved large amounts of nitrogen removal in wastewater treatment due to its long-term properties such as minimal or no oxygen consumption, no external carbon source augmentation, and no CO_2 generation. Furthermore, anammox bacteria's high tolerance for ammonium and nitrogen, organic waste, temperature, and other pollutants including fluorine and phenol has made them a potential wastewater mitigation option. As a result, nitritation/anammox procedures have gained a lot of traction in wastewater mitigation. Numerous researches focused on eliminating contaminants from wastewater, including ammonia nitrogen, COD, total nitrogen, and other pollutants. A key influence on anammox development and nitrite generation in normal settings is the use of anammox for wastewater treatment under adverse conditions, such as a microalgae granular sludge system exposed to UV irradiation and reduced oxygen levels. In Sri Lanka, the ammonia-oxidizing approach was recently established as a positive leachate treatment, with excellent outcomes. Future studies should also include the use of granular anammox on a trial basis (Engr'acia Costa & Julio P´erez, 2006; Hang-Wei & He, 2017).

Conclusive outlook

Ammonia is a significant contaminant of the environment, affecting soil, water, and air. There are various microbial ammonia removal techniques, nitrification and denitrification are two traditional methods for removing ammonia. Nitrification occurs when ammonia-oxidizing bacteria (AOB) and nitric oxide oxidizing bacteria (NOOB) collaborate *chemolithoautotrophically* (NOB). Denitrification occurs after nitrification (NO_3 to N_2), which is carried out by chemo-organo-heterotrophic bacteria in a series of intermediate processes. This is an extremely energy-intensive process that creates various environmentally harmful gases such as N_2O, carbon dioxide, and other nitrogen oxides. Since its invention, the anammox process has piqued the interest of many scientists who would like to explore more about how it may be used to treat NH_3-rich wastewater. Anammox is a one-step process that involves a specific bacterium assisting in the translation of ammonia to nitrogen under anoxic circumstances without the use of a peripheral carbon source. Anammox's employment in water treatment facilities is limited by its measured growth, unpredictability under fluctuating intake, and the availability of NO_2. In comparison to classical nitrifiers, in a single organism, Comammox can complete full nitrification (NH_3 to NO_3), which is more energy-efficient. Furthermore, because anammox uses NO_2 for ammonia oxidation, NO_2 buildup may be prevented. This process also aids in the rapid enrichment of bacteria engaged in wastewater ammonia reduction. This chapter covers a broad range of hot button issues, such as the sources of additional organic pollutants in wastewater, the biochemical and molecular structure and physiological processes associated with anammox and comammox bacteria, the similarities between the anammox and natural nitrogen cycles, and their impact on wastewater ammonia oxidation.

References

Ali, M., Oshiki, M., Awata, T., Isobe, K., Kimura, Z., Yoshikawa, H., Hira, D., Kindaichi, T., Satoh, H., Fujii, T., & Okabe, S. (2015). Physiological characterization of anaerobic ammonium oxidizing bacterium 'C andidatus Jettenia caeni.' *Environmental Microbiology, 17*(6), 2172–2189. https://doi.org/10.1111/1462-2920.12674

Ali, M., & Satoshi, O. (2015). Anammox-based technologies for nitrogen removal: Advances in process start-up and remaining issues. *Chemosphere, 141*, 144–153. https://doi.org/10.1016/j.chemosphere.2015.06.094

Ali, M., Chai, L.-Y., Tang, C.-J., Zheng, P., Min, X.-B., Yang, Z. H., Xiong, L., & Song, Y.-X. (2013) The increasing interest of ANAMMOX research in China: bacteria, process development, and application. *BioMed Research International*. https://doi.org/10.1155/2013/134914

Ali, M., Dario, R. S., Mads, A., & Saikaly, P. E. (2020). Comparative genome-centric analysis of freshwater and marine ANAMMOX cultures suggests functional redundancy in nitrogen removal processes. *Frontiers in Microbiology, 11*, 1637. https://doi.org/10.3389/fmicb.2020.01637

Annavajhala, M. K., Kapoor, V., Santo-Domingo, J., & Chandran, K. (2018). Comammox functionality identified in diverse engineered biological wastewater treatment systems. *Environmental Science and Technology Letters, 5*(2), 110–116.

Arrojo, B., Mosquera-Corral, A., Campos, J. L., & M´endez, R. (2006) Effects of mechanical stress on Anammox granules in a sequencing batch reactor (SBR). *Journal of Biotechnology, 123*(4), 453–463. https://doi.org/10.1016/j.jbiotec.2005.12.023

Atkins, P. F., Jr., Scherger, D. A., Barnes, R. A., & Evans III, F. L. (1973) Ammonia removal by physical-chemical treatment. *Journal* (Water Poll. Control Fed.), 2372–2388. https://www.jstor.org/stable/25038041

Bartlett, R., Mortimer, R. J. G., Morris, K. (2008). Anoxic nitrification: evidence from Humber Estuary sediments (UK). *Chemical Geology, 250*(1–4), 29–39.

Cardemil, C. V., Smulski, D. R., & LaRossa, R. A. (2010). Amy Cheng Vollmer, Bioluminescent Escherichia coli strains for the quantitative detection of phosphate and ammonia in coastal and suburban watersheds. *DNA and Cell Biology, 29*(9), 519–531. https://doi.org/10.1089/dna.2009.0984

Chen, G., Li, J., Wang, Y., Deng, H., Zhang, Y., & Zeng, J. (2016). Novel anammox reactor start-up method using immobilized particles as biocatalyst and its kinetic characteristics. *Desalination and Water Treatment, 57*, 17291–17299.

Costa, E., P´erez, J., & Kreft, J.-U. (2006). Why is metabolic labour divided in nitrification? *Trends in Microbiology,14*(5), 213–219.

Daims, H., Lebedeva, E. V., Pjevac, P., Han, P., Herbold, C., Albertsen, M., Jehmlich Palatinszky, M. N., Vierheilig, J., Bulaev, A., & Kirkegaard, R. H. (2015). Complete nitrification by Nitrospira bacteria. *Nature 528*(7583), 504–509. https://doi.org/10.1038/nature16461

Dang, H., Chen, R., Wang, L., Guo, L., Chen, P., Tang, Z., Tian, F., Li, S., & Klotz, M. G. (2010). Environmental factors shape sediment anammox bacterial communities in hypernutrified Jiaozhou Bay. *China. Applied and Environmental Microbiology, 76*(21), 7036–7047. https://doi.org/10.1128/AEM.01264-10

Desireddy, S., Sabumon, P. C., & Maliyekkal, S. M. (2018). Anoxic ammonia removal using granulated nanoscale oxyhydroxides of Fe (GNOF) in a SBR. *Journal of Environmental Chemical Engineering, 6*(4), 4273–4281. https://doi.org/10.1016/j.jece.2018.05.033

Desireddy, S., Sabumon, P. C., & Maliyekkal, S. M. (2017). Microbial mediated anoxic nitrification-denitrification in the presence of nanoscale oxides of manganese. *International Biodeterioration and Biodegradation, 119*, 499–510.

Egli, K., Fanger, U., Alvarez, P. J. J., Siegrist, H., van der Meer, J. R., & Zehnder, A. J. B. (2001) Enrichment and characterization of an anammox bacterium from a rotating biological contactor treating ammonium-rich leachate. *Archives of Microbiology, 175*(3), 198–207. https://doi.org/10.1007/s002030100255

Fu, Q., Zheng, B., Zhao, X., Wang, L., & Changming, L. (2012). Ammonia pollution characteristics of centralized drinking water sources in China. *Journal of Environmental Sciences, 24*(10), 1739–1743. https://doi.org/10.1016/S1001-0742(11)61011-5

Hong, Y.-G., Li, M., Cao, H., Ji-Dong G. (2011). Residence of habitat-specific anammox bacteria in the deep-sea subsurface sediments of the South China Sea: analyses of marker gene abundance with physical chemical parameters. *Microbial Ecology, 62*(1), 36–47. https://doi.org/10.1007/s00248-011-9849-0

Hsia, T. H., Feng, Y. J., Ho, C. M., Chou, W. P., & Tseng, S. K. (2008). PVA-alginate immobilized cells for anaerobic ammonium oxidation (anammox) process. *J Ind Microbiol Biotechno, 35*, 721–727.

Hu, M., Tang, Y., & Wang, X. (2014). Immobilized microorganism technology and the application of wastewater treatment. *Advanced Materials Research, 955–959*, 383–386.

Hu, B., Shen, L., Du, P., Zheng, P., Xu, X., & Zeng, J. (2012). The influence of intense chemical pollution on the community composition, diversity and abundance of anammox bacteria in the Jiaojiang Estuary (China). *PLoS One 7*(3), e33826. https://doi.org/10.1371/journal.pone.0033826

Hu, H.-W., & He, J.-Z. (2017) Comammox—a newly discovered nitrification process in the terrestrial nitrogen cycle. Journal of Soils and *Sediments, 17*(12), 2709–2717.

Huff, L., Delos, C., Gallagher, K., & Beaman, J. (2013). *Aquatic life ambient water quality criteria for ammonia-freshwater*. US Environmental Protection Agency.

Isaka, K., Kimura, Y., Matsuura, M., Osaka, T., & Tsuneda, S. (2017). First full-scale nitritation-anammox plant using gel entrapment technology for ammonia plant effluent. *Biochemical Engineering Journal, 122*, 115–122.

Jetten, M. S., Niftrik, L. V., Strous, M., Kartal, B., Keltjens, J. T., & Op den Camp, H. J. (2009). Biochemistry and molecular biology of anammox bacteria. *Critical Reviews in Biochemistry and Molecular Biology, 44*(2–3), 65–84. https://doi.org/10.1080/10409230902722783

Jiang, X.-Y., Cheng, Y.-F., Zhu, W.-Q., Bai, Y.-H., Xu, L.-Z.-J., Wu, X.-Q., & Jin, R.-C. (2018). Effect of chromium on granule-based anammox processes. *Bioresource Technology, 260*, 1–8. Elsevier Ltd.

Kartal, B., Rattray, J., Niftrik, L. A., van de Vossenberg, J., Schmid, M. C., Webb, R. I., Schouten, S., Fuerst, J. A., Damst´e, J. S., Jetten, M. S., & Strous, M. (2007). Candidatus "Anammoxoglobus propionicus" a new propionate oxidizing species of anaerobic ammonium oxidizing bacteria. *Systematic and Applied Microbiology, 30*(1), 39–49. https://doi.org/10.1016/j.syapm.2006.03.004

Leenen, E. J. T. M., Dos Santos, V. A. P., Grolle, K. C. F., Tramper, J., & Wijffels, R. H. (1996). Characteristics of and selection criteria for support materials for cell immobilization in wastewater treatment. *Water Research, 30*, 2985–2996.

Lei, H., Xia, X., Li, C., & Xi, B. (2012). Decomposition analysis of wastewater pollutant discharges in industrial sectors of China (2001–2009) using the LMDI I method. *International Journal of Environmental Research and Public Health, 9*(6), 2226–2240. https://doi.org/10.3390/ijerph9062226

Li, X., & Sung, S. (2015). Development of the combined nitritation–anammox process in an upflow anaerobic sludge blanket (UASB) reactor with anammox granules. *Chemical Engineering Journal, 281*, 837–843. https://doi.org/10.1016/j.cej.2015.07.016

Li, M., & Gu, J.-D. (2011). Advances in methods for detection of anaerobic ammonium oxidizing (anammox) bacteria. *Applied Microbiology and Biotechnology, 90*(4), 1241–1252. https://doi.org/10.1007/s00253-011-3230-6

Li, J., Li, J., Peng, Y., Wang, S., Zhang, L., Yang, S., & Shuai, L. (2020). Insight into the impacts of organics on anammox and their potential linking to system performance of sewage partial nitrification-anammox (PN/A): a critical review, *Bioresource Technology, 300*, 122655. https://doi.org/10.1016/j.biortech.2019.122655

Li, H., Chen, S., Mu, B.-Z., & Ji-Dong, G. (2010). Molecular detection of anaerobic ammonium-oxidizing (anammox) bacteria in high-temperature petroleum reservoirs. *Microbial Ecology, 60*(4), 771–783. https://doi.org/10.1007/s00248-010-9733-3

Li, M., Hong, Y., Klotz, M. G., & Ji-Dong, G. (2010). A comparison of primer sets for detecting 16S rRNA and hydrazine oxidoreductase genes of anaerobic ammonium-oxidizing bacteria in marine sediments. *Applied Microbiology and Biotechnology, 86*(2), 781–790. https://doi.org/10.1007/s00253-009-2361-5

Li, J., Li, J., Gao, R., Wang, M., Yang, L., Wang, X., Zhang, L., Peng, Y. (2018). A critical review of one-stage anammox processes for treating industrial wastewater: optimization strategies based on key functional microorganisms, *Bioresource Technology, 265*, 498–505. https://doi.org/10.1016/j.biortech.2018.07.013

Liu, M., Huang, X., Song, Y., Tang, J., Cao, J., Zhang, X., Zhang, Q., Wang, S., Xu, T., Kang, L., & Cai, X. (2019a). Ammonia emission control in China would mitigate haze pollution and nitrogen deposition, but worsen acid rain. *Proceedings of the National Academy of Sciences, 116*(16), 7760–7765. https://doi.org/10.1073/pnas.1814880116

Liu, Y., Ngo, H. H., Guo, W., Peng, L., Wang, D., & Ni, B. (2019b). The roles of free ammonia (FA) in biological wastewater treatment processes: A review. *Environment International,123*, 10–19.

Magrí, A., Vanotti, M. B., & Szögi, A. A. (2012). Anammox sludge immobilized in polyvinyl alcohol (PVA) cryogel carriers. *Bioresource Technology, 114*, 231–240.

Mardanov, A. V., Beletsky, A. V., Ravin, N. V., Botchkova, E. A., Litti, Y. V., & Nozhevnikova, A. N. (2019). Genome of a novel bacterium "Candidatus jettenia ecosi" reconstructed from the metagenome of an anammox bioreactor. *Frontiers in Microbiology, 10*, 2442. https://doi.org/10.3389/fmicb.2019.02442

Morris, S., Garcia-Cabellos, G., Ryan, D., Enright, D., & Enright, A. (2019). Low-cost physico-chemical treatment for removal of ammonia, phosphate and nitrate contaminants from landfill leachate. *Journal of Environmental Science and Health, Part A, 54*(12), 1233–1244. https://doi.org/10.1080/10934529.2019.1633855

Mulder, A., Van de Graaf, A. A., Robertson, L. A., & Kuenen, J. G. (1995) Anaerobic ammonium oxidation discovered in a denitrifying fluidized bed reactor. *FEMS Microbiology Ecology, 16*(3), 177–183. https://doi.org/10.1111/j.1574-6941.1995.tb00281.x

Okamoto, H., Kawamura, K., Nishiyama, T., Fujii, T., & Furukawa, K. (2013) Development of a fixed-bed anammox reactor with high treatment potential. *Biodegradation 24*(1), 99–110. https://doi.org/10.1007/s10532-012-9561-x

Østgaard, K., Wangen, B. F., Knutsen, S. H., & Aasen, I. M. (1993). Large-scale production and purification of kcarrageenase from Pseudomonas carrageenovora for applications in seaweed biotechnology. *Enzyme and Microbial Technology, 15*, 326–333.

Park, T.-J., Lee, J.-H., Lee, M.-S., Park, C.-H., Lee, C. H., Moon, S.-D., Jiwoong, C., Rongxue, C., An Youn, J., Dong, H. Y., Lee, S. H., Lee, J.-K., & Kyung, D. Z. (2018). Development of water quality criteria of ammonia for protecting aquatic life in freshwater using species sensitivity distribution method. *Science of the Total Environment, 634*, 934–940. https://doi.org/10.1016/j.scitotenv.2018.04.018

Rossi, S., Díez-Montero, R., Rueda, E., Castillo Cascino, F., Parati, K., García, J., & Ficara, E. (2020). Free ammonia inhibition in microalgae and cyanobacteria grown in wastewaters: Photo-respirometric evaluation and modelling. *Bioresource Technology, 305*, 123046.

Seung, H. S., Suk, S. C., Park, K., & Yoo, Y. J. (2005). Novel hybrid immobilization of microorganisms and its applications to biological denitrification. *Enzyme and Microbial Technology, 37*, 567–573.

Shah, M. P. (2020). *Microbial bioremediation and biodegradation*. Springer.

Shah, M. P. (2021). *Removal of refractory pollutants from wastewater treatment plants*. CRC Press.

Sumino, T., Nakamura, H., Mori, N., & Kawaguchi, Y. (1992). Immobilization of nitrifying bacteria by polyethylene glycol prepolymer. *Journal of Fermentation and Bioengineering, 73*, 37–42.

Swathi Desireddy, P. C., Sabumon, R. L., & Manasa, A. M. (2020) Development of an up-flow anoxic nano-biotechnological reactor for simultaneous removal of ammonia and COD from low

C/N secondary treated wastewater. *Journal of Water Process Engineering, 36*, 101344. https://doi.org/10.1016/j.jwpe.2020.101344.

Takeshima, M., Kimata, T., Moil, N., & Emori, H. (1993). "Pegasus", an innovative high-rate BOD and nitrogen removal process for municipal wastewater. In *66th WEF annual conference.*

Thu, B., Smidsrød, O., & Skjak-Brk, G. (1996). Alginate gels—some structure-function correlations relevant to their use as immobilization matrix for cells. *Program Biotechnology*, 19–30.

Van Teeseling, C. F., Muriel, N. M., de Almeida, A. K., Daan R. S., Huub J. M., den Camp, O., Rachel, R., Jetten, M. S. M., & Niftrik, L. (2014). A new addition to the cell plan of anammox bacteria: "Candidatus Kuenenia stuttgartiensis" has a protein surface layer as the outermost layer of the cell. *Journal of Bacteriology, 196*(1), 80–89. https://doi.org/10.1128/JB.00988-13

Viancelli, A., Kunz, A., Esteves, P. A., Bauermann, F. V., Furukawa, K., Fujii, T., Antonio, R. V., & Vanotti, M. (2011). Bacterial biodiversity from an anaerobic up flow bioreactor with ANAMMOX activity inoculated with swine sludge. *Brazilian Archives of Biology and Technology, 54*(5), 1035–1041. https://doi.org/10.1590/S1516-89132011000500022

Wang, Y., Zhu, G., Harhangi, H. R., Zhu, B., Jetten, M. S. M., Yin, C., & Op den Camp, H. J. M. (2012) Co-occurrence and distribution of nitrite-dependent anaerobic ammonium and methane-oxidizing bacteria in a paddy soil. *FEMS Microbiology Letters, 336*(2), 79–88, https://doi.org/10.1111/j.1574-6968.2012.02654.x

Wang, B., Peng, Y. Z., Guo, Y., Zhao, M., & Wang, S. (2016a). Illumina MiSeq sequencing reveals the key microorganisms involved in partial nitritation followed by simultaneous sludge fermentation, denitrification and anammox process, *Bioresource Technology, 207*, 118–125. https://doi.org/10.1016/j.biortech.2016.01.072

Wang, T., Shen, B., Zhang, S., Wang, Z., & Tian, L. (2016b). Start-up performance of Anammox process in a fixed bed reactor (FBR) filled with honeycomb-like polypropylene carriers. *Water Science and Technology, 73*(8), 1848–1854. https://doi.org/10.2166/wst.2016.017

Yao, J., Xia, G., Mei, Y., Lu, Y., Xu, D., Sun, N., Wang, J., & Chen, J. (2019). Process optimization of electrochemical oxidation of ammonia to nitrogen for actual dyeing wastewater treatment. *International Journal of Environmental Research and Public Health, 16*(16), 2931. https://doi.org/10.3390/ijerph16162931

Zhang, L., Zheng, P., Tang, C.-J., & Ren-Cun, J. (2020). Anaerobic ammonium oxidation for treatment of ammonium-rich wastewaters. *Journal of Zhejiang University Science B, 9*(5), 416–426.https://doi.org/10.1631/jzus.B0710590

Zhang, L., & Satoshi, O. (2020). Ecological niche differentiation among anammox bacteria. *Water Research, 171*, 115468. https://doi.org/10.1016/jwatres.2020.115468

MIX
Papier aus verantwortungsvollen Quellen
Paper from responsible sources
FSC® C105338

If you have any concerns about our products,
you can contact us on
ProductSafety@springernature.com

In case Publisher is established outside the EU,
the EU authorized representative is:
Springer Nature Customer Service Center GmbH
Europaplatz 3, 69115 Heidelberg, Germany

Printed by Libri Plureos GmbH
in Hamburg, Germany